# Biochemistry Research Trends

nova
science publishers

www.novapublishers.com

# Biochemistry Research Trends

**The Medical Biology Guide to Proteins**
David Aebisher, PhD, DSc (Editor)
2023. ISBN: 979-8-88697-910-7 (Softcover)
2023. ISBN: 979-8-89113-024-1 (eBook)

**Water in Biology: A Molecular View**
Michael E. Green, PhD (Editor)
Alisher M Kariev (Editor)
2023. ISBN: 979-8-88697-708-0 (Hardcover)
2023. ISBN: 979-8-88697-754-7 (eBook)

**Biomolecules and Corrosion**
Santosh Kumar Karn, PhD (Editor)
Anne Bhambri (Editor)
2023. ISBN: 979-8-88697-458-4 (Softcover)
2023. ISBN: 979-8-88697-531-4 (eBook)

**The Biochemical Guide to Enzymes**
David Aebisher, PhD (Editor)
Dorota Bartusik-Aebisher, PhD (Editor)
2022. ISBN: 979-8-88697-410-2 (Softcover)
2022. ISBN: 979-8-88697-518-5 (eBook)

**Mineral Water: From Basic Research to Clinical Applications**
Maria João Martins, PhD (Editor)
2022. ISBN: 978-1-68507-458-6 (Hardcover)
2022. ISBN: 978-1-68507-541-5 (eBook)

More information about this series can be found at
https://novapublishers.com/product-category/series/biochemistry-research-trends/

**David Aebisher**
Editor

# The Medical Biology Guide to Proteins

nova science publishers

www.novapublishers.com

DOI: https://doi.org/10.52305/WXOF4365



## NOTICE TO THE READER

The Publisher has taken reasonable care in the preparation of this book but makes no expressed or implied warranty of any kind and assumes no responsibility for any errors or omissions. No liability is assumed for incidental or consequential damages in connection with or arising out of information contained in this book. The Publisher shall not be liable for any special, consequential, or exemplary damages resulting, in whole or in part, from the readers' use of, or reliance upon, this material. Any parts of this book based on government reports are so indicated and copyright is claimed for those parts to the extent applicable to compilations of such works.

Independent verification should be sought for any data, advice or recommendations contained in this book. In addition, no responsibility is assumed by the Publisher for any injury and/or damage to persons or property arising from any methods, products, instructions, ideas or otherwise contained in this publication.

This publication is designed to provide accurate and authoritative information with regards to the subject matter covered herein. It is sold with the clear understanding that the Publisher is not engaged in rendering legal or any other professional services. If legal or any other expert assistance is required, the services of a competent person should be sought. FROM A DECLARATION OF PARTICIPANTS JOINTLY ADOPTED BY A COMMITTEE OF THE AMERICAN BAR ASSOCIATION AND A COMMITTEE OF PUBLISHERS.

## Library of Congress Cataloging-in-Publication Data

Names: Aebisher, David, editor.
Title: The medical biology guide to proteins / David Aebisher (editor)
Description: New York : Nova Science Publishers, [2023] | Series: Biochemistry research trends | Includes bibliographical references and index. |
Identifiers: LCCN 2023029954 (print) | LCCN 2023029955 (ebook) | ISBN 9798886979107 (paperback) | ISBN 9798891130241 (adobe pdf)
Subjects: LCSH: Proteins. | Human physiology. | Medicine.
Classification: LCC QP551 .M3695 2023 (print) | LCC QP551 (ebook) | DDC 572/.6--dc23/eng/20230724
LC record available at https://lccn.loc.gov/2023029954
LC ebook record available at https://lccn.loc.gov/2023029955

Published by Nova Science Publishers, Inc. † New York

# Contents

# Preface

This book aims to provide scientific information about proteins. 28 chapters have been written that describe the characteristics of proteins and their influence on human physiology. The authors presented the structure of individual proteins and their role in particular diseases. Proteins are among the most important nutrients responsible for the proper functioning of the body. Protein is one of the basic building blocks of our tissues and plays a role in transport, regulating biochemical processes and in reactions initiated by the immune system.

Chapter 1 - The NMP-22 protein is a commonly used protein in the diagnostics of bladder cancer. The sequence of its individual amino acids is encoded on the 1q25 chromosome, its canonical length is 2115. It regulates the separation of chromatids and daughter cells. The NMP-22 protein is a protein that is involved in controlling the process of cell multiplication. Its main application is to be used as a marker protein for the detection of neoplastic cells in bladder cancer. Tumors can be detected at an early stage due to the high sensitivity of NMP-22. The research on the NMP-22 protein is still ongoing and shows that is a marker of neoplastic changes in the bladder. However, work on the NMP-22 protein is still ongoing and in the future it can be used in the routine examination of the urinary bladder.

Chapter 2 - Carbohydrate antigen 15.3 or cancer antigen 15.3, (CA 15.3) is one of the most commonly used tumor markers. As many studies have shown, CA 15.3 is an invaluable biological marker. It is widely used in oncological diagnostics, but also in pulmonology. The 15.3 antigen is used to detect neoplastic diseases (breast cancer, ovarian cancer or lung cancer) and lung fibrosis, systemic sclerosis and rheumatoid arthritis. Disease prevention is as important as diagnosis. Contrary to appearances, the most important to perform regular preventive examinations adapted to the age and health condition of the patient (annual blood tests, and in women: cytology, ultrasound of breasts and reproductive organs). Genetic testing may be helpful if there has been a previous history of cancer in the patient's family.

Chapter 3 - Thyroglobulin is a substance classified as relatively large glycoprotein with a mass of 660 kDa. Functionally, it is a protein hormone synthesized by the epithelial cells of the thyroid gland, which, further secreted into the lumen of the follicles of this organ. It is estimated that up to 5% of the total number of people on Earth suffer from diseases related to the malfunctioning of the thyroid gland. This gland is especially important for the proper development of the body. The secretion products of this gland affect the operation of almost all systems of the human body. After oncological treatment and total thyroidectomy, the concentration of these antibodies should decrease, but if they are not, and their level increases, it indicates an increased risk of persistent cancer or its recurrence. Further research on this fundamental molecule of the thyroid gland may lead to the development of innovative methods enabling a more precise assessment of the condition of many patients, especially those struggling with thyroid abnormalities or its neoplasms.

Chapter 4 - Carcinoembryonic antigen (CEA) is one of the most used tumor markers in the world. CEA is a group of foetal glycoproteins produced in the developing tissues of the foetal digestive tract, pancreas and liver. Their synthesis is almost completely stopped before birth and in healthy adults they are not produced in significant amounts, therefore the level of CEA in the blood is very low, less than 5ng / ml. CEA can be designated as a non-specific cancer marker. The discovery of this antigen initiated a series of studies enabling the use of the characteristics of these glycoproteins in the diagnosis and monitoring of the course of malignant colon cancer.

Chapter 5 - Carbohydrate antigen 125 (CA-125), also known as MUC16, is a high molecular weight soluble glycoprotein belonging to the mucin family (MUC). CA-125 was first discovered in an ovarian cancer cell line. There are also hypotheses that it is also secreted by tissues of neoplastic origin. CA-125 has long been used as a marker for ovarian tumors. Elevated serum levels of CA-125 may also indicate other neoplasms such as lung and mediastinal cancer, teratoma or non-Hodgkin's lymphoma. Monitoring the concentration of carbohydrate antigen 125 in the blood makes it possible to predict the likelihood of this cancer developing. Long-term control of CA125 concentration enables segregation of patients into groups with low, medium and high risk of ovarian cancer development, which is of great importance in the early detection of the disease. Due to the high frequency of this disease, it is important to further develop research on the carbohydrate antigen 125.

Chapter 6 - Calcitonin (CT) is an animal polypeptide hormone that consists of 32 amino acids. The mass of a single molecule is 3.42 kDa. The basic role of the hormone is to regulate the phosphorus and calcium balance of the body. For a long time in the history of calcitonin research, the role of this hormone has given rise to scientific debate about its exact action and effect on living organisms. This was mainly due to the fact that its deficiency or excess was not observed in patients with diseases of the skeletal system, although its physiological effects were documented. The exact role it plays in human physiology is not yet fully understood. In laboratory studies, the most important reason for determining its level is still the study of medullary thyroid cancer (MCT), a disease for which calcitonin is a sensitive and specific marker.

Chapter 7 - Epidermal growth factor (EGF) is a small protein, composed of 53 amino acids and completely devoid of alanyl, phenylalanyl, or lysine residues. Its structure includes 3 internal disulfide bonds, and the hormone itself has a second-order structure, periodically assuming the form of a β-sheet. Extremely important protein needed to maintain homeostasis in the human body. It plays a fundamental role in stimulating epithelial cells to divide. Due to this property, intensive research has been carried out to determine the use of EGF in the wound healing process. The effect of using these antibodies is the increased sensitivity of cancer cells to radiotherapy, inhibition of tumor growth and induction of apoptosis processes in glioblastoma multiforme cells. Currently, laboratories around the world are working intensively to better understand the role of this factor in the human body, and to discover new, previously unknown properties that may significantly affect our lives in the future.

Chapter 8 - Estrogen and progesterone are sex hormones synthesized from cholesterol and therefore they belong to the steroid hormones. Estrogen and progesterone regulate the body's work by acting on various types of receptors. They are most often associated with the female reproductive cycle because they are essential for its proper course. Research on individual types of receptors is very promising, as these receptors could be used to prevent and inhibit many pathologies, such as endometriosis or cancers of the reproductive system.

Chapter 9 - Xanthine oxidase is the major protein involved in the formation of uric acid during purine metabolism. Its deficiency or excess is noticeable thanks to the determination of the concentration of purine. Xanthine oxidase is often found in the lungs, liver and serum. It plays a very important role in the metabolism of purines, and additionally catalyzes the reactions of caffeine metabolism, drugs, and the biosynthesis of second-order metabolites. It also significantly influences the process of oxidative stress. This study presents the negative effects of this protein by reducing the amount of reactive oxygen species. This research presents molecules with properties to reduce the number of ROS by binding them or eliminating their negative effects, but not acting directly on the reaction of uric acid formation.

Chapter 10 - The presented research shows the potential of selectins in the diagnosis of many diseases. Scientists conducting research and then publishing them in articles focused in particular on diabetes, cancer, atherosclerosis, and myocardial infarction. In their hypotheses, they assumed that there was a relationship between the concentration of selectins in the organisms of patients and their diseases. In recent years, researchers have published few articles on selectins.

Chapter 11 - Integrins are a group of heterodimers in vertebrates consisting of 24 different transmembrane receptors. Integrins are the most important group among cell adhesion receptors, and also one of the most numerous groups of external membrane receptors. They are present virtually in every cell of the body, except for red blood cells. The main focus of this chapter was on their structural and biochemical features. The studies were mainly conducted using antibodies and the ligand binding mechanism in terms of mutations. In the cellular signals still hide many unknowns.

Chapter 12 - Coronins are conserved, low-evolving proteins that play an important role in the immune system. Members of the coronin protein family are important regulators of the actin cytoskeleton and are related to actin. Coronins functions have not yet been fully described. It is known, however, that these proteins are important in the functioning of immunity, especially natural immunity.

Chapter 13 - Ependimin is a glycoprotein that was first discovered in osseous fish in the lining of the cavities of the central nervous system. The significant amount of ependymin is also found in the cerebrospinal and extracellular fluid in the brain of these animals. Ependimin is related to EPDR glycoproteins with a variety of functions. The structures of EPDR are similar in all organisms in which it occurs. The monomeric subunit consists of two antiparallel planes and a hydrophobic pocket, probably used for lipid attachment. Studies have been carried out to investigate the influence of the EPDR1 gene on the development of neoplasms: HCC, BLCA and CRC. It has been shown that in neoplastic cells the expression of this gene is higher than in healthy cells. The role of EPDR1 in the regulation of cancer resistance is related to the movement of immune cells to the vicinity of the tumor. It seems reasonable to conduct research on the use of the EPDR1 gene in medicine as a biomarker for cancer.

Chapter 14 - Cadherins are transmembrane proteins involved in the formation of adhesive bonds between cells. Protein plays a key role in several important processes during embryogenesis, such as the formation of gastrula, neurula, and organogenesis. Adhesion based on this protein is necessary to maintain the proper architecture of tissues in organisms. Despite the fact that cadherins have been discovered for over 40 years, their development is still slow and requires significant financial outlays. However, their incorrect expression leads to oncogenesis and metastasis. Therefore, a better understanding of cadherins is critical to cancer clinical applications, especially as therapeutic targets. Understanding how N-cadherin influences cell behavior will enable the development of therapies to combat its activity and prevent cancer cell growth, invasion and metastasis. Adhesion molecules are promising new targets in the treatment of cancer but could also be useful in predicting patient prognosis in human and veterinary medicine.

Chapter 15 - Actin is one of the most important proteins that build the cytoskeleton and nucleoskeleton of a eukaryotic cell. It is a conserved protein, the most abundant in the cell, and it has the ability to polymerize and forms complexes with other proteins. It can undergo mutations in human organisms, usually they are missense mutations, causing systemic diseases. It correlates with a very large number of other proteins. In the human body, it is coded by 6 genes and has many important functions that enable it to function properly. Very important function performed by nuclear actin is the structure of the nucleoskeleton and the control of transcription. It was first discovered in the 19th century.

Chapter 16 - Natural representatives of opioids group are endorphins, dynorphins and enkephalins. The synthetic representatives of opioids group are fentanyl, pethidine and methadone. They can be found in the neuroendocrine system, spinal cord, brain, intestinal and peripheral nervous systems, and in the endocrine part of the pancreas.

Chapter 17 - MAP17 is a small (17 kDa), non-glycosylated membrane protein located in the plasma membrane and the Golgi apparatus. The physiological role of this protein in the proximal tubules is not well understood, however, MAP17 stimulates SGLT transporters, increasing the specific Na-dependent transport of mannose and glucose in oocytes and human tumor cells. MAP17 is overexpressed, mainly through mRNA amplification, in various human cancers. Through the enhanced tumorigenic properties induced by MAP17, they are associated with an increase in ROS, since MAP17 significantly alters the mRNA levels of genes involved in oxidative stress and increases endogenous ROS, and antioxidant treatment of

MAP17-expressing cells reduces their cell carcinogenic properties. Generalized overexpression of MAP17 in human cancers indicates that MAP17 may be a good marker of neoplasm, especially malignant progression.

Chapter 18 - Erythropoietin (EPO) is a glycoprotein hormone that is naturally produced by the peritubular cells of the kidney to stimulate the production of red blood cells. Literature results confirmed the presence of Epo and EpoR in malignant tumors of the larynx and showed a correlation between Epo expression and survival. Hypoxia is the primary stimulus for the production of erythropoietin, which acts through the erythropoietin receptor. As a result of hypoxia within the tumor, the transcription factor HIF, and in particular HIF-1α, induces the production of the glycoprotein hormone erythropoietin (EPO) in kidney and liver cells. The concentration of EPO and EPO-R in cancers of the oral cavity, pharynx and larynx is significantly increased compared to healthy tissues. The accurate diagnosis and subsequent prediction of the course of a malignant larynx disease remains a challenge.

Chapter 19 - PD-L1 is a validated biomarker used to guide treatment choice in clinical practice. Over the past 10 years, cancer immunotherapy has made significant advances in many cancers and is gradually being applied in clinical oncology care, including Programmed Cell Death Protein-1 (PD-1)/Programmed Cell Death Ligand 1 (PD-L1 Access). Compared to traditional therapies, the emerging PD-1/PD-L1 blocking immunotherapy shows more satisfactory therapeutic effects and less toxicity in patients with advanced squamous cell carcinoma of the head and neck. Expression of PD-L1 on cells of the immune system in a tumor biopsy before treatment indicates a previously induced adaptive anti-tumor immune response and is associated with improved treatment outcomes. Understanding the complex mechanism behind PD-L1 presentation in TME may enable therapeutic approaches to regulate the expression of this immunosuppressive ligand to enhance the PD-1 blockade effect.

Chapter 20 - Although serotonin is produced by a very small number of neurons, it has countless functions in the human body. By affecting the digestive, circulatory, endocrine, urogenital and central nervous systems, it controls such processes as respiration, metabolism, digestion, heart function, vascular contractility, blood homeostasis, micturition, reproduction and behavioral processes. The possible use of excessive or reduced stimulation of serotonin receptors in the treatment of many diseases is at the stage of research. Due to differences in structure, function, action and ligands, 5-HT receptors have been divided into 7 families and at least 15 variants. Most of them belong to the receptors associated with the G protein, 5-HT3 belongs to the ionic receptors.

Chapter 21 - GABA receptors are a group of membrane receptors that are widely distributed not only within the central nervous system but also in other organs. Based on their structural structure, GABA receptors have been divided into 4 classes: A, B, C and F. Each of them is characterized by a different structure, which determines its pharmacokinetic properties, the degree of binding with various ligands and the functions of the receptor. The GABAergic system has a number of functions. The influence of GABA receptors begins in fetal life, where they ensure the correct process of shaping neural networks. Later, they watch over such processes as learning, memory, and the proper shaping of emotions and behavior. They play an important role in the pathomechanism of many CNS diseases.

Chapter 22 - Bacterial ADP-ribosylation toxins constitute a large family of dangerous toxins, including pertussis, cholera and diphtheria toxins, which, as cytotoxic agents, cause severe infectious diseases, including whooping cough, cholera, and diphtheria. Among these toxins, Pectenotoxin (PTX) is the predominant. It catalyses ADP ribosylation of the α-subunits of the Gi/o family of heterotrimeric proteins (Gαi, Gαo and Gαt), thus preventing interaction of the G proteins with their cognate G protein coupled receptors (GPCR). PTX plays a key role in the pathogenesis of whooping cough, the development of protective immunity against reinfection, and is an essential component of new acellular vaccines.

Chapter 23 - The estrogen receptors have many important functions during the development and maturation of tissues. Their expression is not only related to the reproductive system, but also occurs in the lungs, prostate, cardiovascular and nervous systems. The results of many studies carried out on their specification and mechanism of action have made it possible to create many therapies for various diseases in which estrogen receptors are involved in their formation.

Chapter 24 - Ricin is a potent toxin derived from the castor plant, Ricinus communis L. Ricin intoxication mimics a variety of disease states, thus a low threshold of suspicion must be maintained to recognize a potential epidemic. The castor plant, native to the southeastern Mediterranean region, eastern Africa, and India, it is now widespread throughout temperate and subtropical regions. Experimental animal studies reveal that clinical signs and pathological manifestations of ricin toxicity depend on the dose as well as the route of exposure. Contact with ricin powders or products may cause redness and pain of the skin and the eyes. Underway to develop small molecule inhibitors for the treatment of ricin intoxication. Recent findings suggest that refinement of the newly identified ricin inhibitors will yield improved compounds suitable for continued evaluation in clinical trials.

Chapter 25 - The nicotinic acetylcholine receptors belong to the group of acetylcholine responsive polypeptide receptors. They are ligand-gated ion channels and can be divided into two groups: muscle receptors, which are found at the neuromuscular junction of skeletal muscles where they mediate neuromuscular transmission, and neuronal receptors, which are found throughout the peripheral and central nervous systems. In the peripheral nervous system, they transmit signals from presynaptic cells to postsynaptic cells in the sympathetic and parasympathetic nervous systems. A single receptor is pentametric around a water-filled pore. The compounds with the ability to activate prescription acetylcholine include, among others nicotine, epibatidine, anabasein, α anotoxin, arecoline, lobeline or cytisine. Activation of nicotinic receptors leads to two main mechanisms. The first is the flow of cations through the receptor which depolarizes the cell membrane resulting in an excitatory postsynaptic potential in neurons, leading to the activation of voltage-gated ion channels. The second effect is the influence of Ca2+ ions, which indirectly affects various intracellular cascades, which leads to the regulation of the activity of certain genes or the release of neurotransmitters.

Chapter 26 - Thyroglobulin (TG) is a protein produced by the thyroid gland, more specifically by thyroid follicular cells. Its production is stimulated by intra-thyroid deficiency or excess of iodine and by the presence of immunoglobulins that stimulate the functioning of the thyroid gland. The serum TG level post-surgery reflects the amount of residual thyroid mass. Serum thyroglobulin measurement is essential in the diagnosis and follow-up of several thyroid disorders. An increasing TG concentration, when on a suppressive dose of thyroxine, indicates the recurrence of tumor or metastatic spread. Thyroglobulin assays are now in widespread use as a tumor marker for monitoring patients with differentiated thyroid carcinoma.

Chapter 27 - Threonine is an organic chemical compound, an electrically neutral amino acid. Its full name is α-Amino-β-Hydroxybutyric Acid, and the chemical formula is C4H9NO3. It belongs to the exogenous amino acids, which means that the body is not able to produce it on its own, and at the same time it is essential and must be supplied with food.

Chapter 28 - Lysine (l-lysine; abbreviated name Lys, single letter abbreviation K) is an organic chemical compound belonging to the group of protein amino acids. For humans, it is an exogenous amino acid, i.e., it is not synthesized in the body and should be supplied with food.

# Chapter 1

# Nuclear Matrix Protein 22 (NMP-22)

**Katarzyna Wąchała**
**Dorota Bartusik-Aebisher***
**and David Aebisher**
Medical College of the University of Rzeszów, Poland

## Abstract

The NMP-22 protein is a commonly used protein in the diagnostics of bladder cancer. The sequence of its individual amino acids is encoded on the 1q25 chromosome, its canonical length is 2115. It regulates the separation of chromatids and daughter cells. The NMP-22 protein is a protein that is involved in controlling the process of cell multiplication. Its main application is to be used as a marker protein for the detection of neoplastic cells in bladder cancer. Tumors can be detected at an early stage due to the high sensitivity of NMP-22. The research on the NMP-22 protein is still ongoing and shows that is a marker of neoplastic changes in the bladder. However, work on the NMP-22 protein is still ongoing and in the future it can be used in the routine examination of the urinary bladder.

**Keywords:** nuclear matrix protein 22 (NMP-22), Food and Drug Administration (FDA), bladder cancer, tumors, protein

The NMP-22 protein is a commonly used protein in the diagnosis of bladder cancer. The sequence of its individual amino acids is encoded on the 1q25

* Corresponding Author's Email: dbartusikaebisher@ur.edu.pl.

In: The Medical Biology Guide to Proteins
Editor: David Aebisher
ISBN: 979-8-88697-910-7

chromosome. In humans, NMP-22 is present in the cell nucleus of the body's cells. It is a protein that, during mitosis, is located at the opposite poles of the karyokinetic spindle. It regulates the separation of chromatids and daughter cells. Intensive research on this protein began in 2002, when the American Food and Drug Administration (FDA) approved its introduction to the market (Costantini et al., 2021).

The NMP-22 protein is very much a protein that is involved in controlling the process of cell multiplication. Its amount in a cell is related to its metabolic activity and its ability to divide. Most of this protein is found in the cell nucleus. It forms the internal scaffold of the cell nucleus, is also associated with DNA replication and RNA synthesis, and regulates the mitosis process. Its amount increases significantly during cell division. Bladder cancer is one of the more common cancers that has a high recurrence rate, and a patient who has suffered from it requires continued clinical surveillance for life through regular Pap smears. NMP-22 is a marker of urothelial cell death and is elevated in the urine of patients with bladder cancer, which is why the NMP-22 protein plays a significant role and is used in tests to detect bladder tumors as well as very early stages of cancer (Miyake et al., 2017). This is possible due to their high sensitivity. This has led to the fact that research methods using this test are many times more effective and have better diagnostic results than cytology. Unfortunately, due to the insufficient number of experimental studies, cystoscopy cannot be replaced in the observation of neoplastic changes in the urinary bladder. NMP-22 may be present in small amounts in the urine of healthy people. They can be found in much larger amounts in people suffering from cancer originating in the transitional epithelium, otherwise known as the urinary tract epithelium (Mati et al., 2020). This test works by detecting the presence of NMP-22 in the urine, an enzyme-linked immunosorbent reaction using two types of monoclonal antibodies and an appropriate enzyme, which involves the quantitative determination of the Nuclear Matrix Protein in the urine. The sensitivity of this test is influenced by the size of the patient's tumors and its advancement, the greater the changes, the greater the sensitivity of the tests. As a result, this method is much less able to detect recurrent tumors, as they are usually smaller in size and volume. Very often in diagnostics we can also meet with false positive results, most often it is due to the fact that the NMP-22 protein is released from cells during their apoptosis, which may be the result of inflammation of the urinary bladder, which very often manifests itself e.g., hematuria (Sajid et al., 2020). Currently, these tests are used as auxiliary tests for diagnostic tests. There is still a large amount of scientific work in progress to test the effectiveness of

NMP-22 as a bladder cancer marker. The first work on it dates back to 1998. In the last 5 years, as many as 12 scientific papers on this protein have been documented, which only shows how much interest scientists and researchers enjoy. The largest amount of research on this protein was carried out in the years 2000-2005, during this period more than 20 scientific papers were published, mainly on the topic of using the NMP-22 protein as a marker of bladder cancer (Sajid et al., 2020).

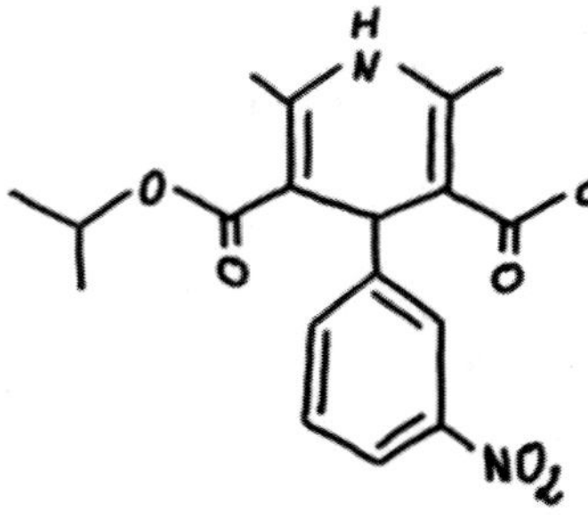


**Figure 1.** NMP protein chemical structure.

In summary, the NMP-22 protein is one of the most important proteins that was discovered in the 20th century. Its main application is its use as a marker protein for the detection of neoplastic cells in bladder cancer. Tumors can be detected at an early stage due to the high sensitivity of NMP-22. It is worth noting that its effectiveness and diagnostic value is significantly greater than in the case of cytological tests. All this is possible thanks to the presence of a large amount of this protein in the epithelium of the urinary tract of people suffering from bladder cancer. This is due to the natural presence of this protein in the nucleus of eukaryotic cells, the amount of which increases significantly during numerous, extra and uncontrolled cell divisions. In the urine, the NMP-22 proteins are detected in the urine by means of an enzyme-linked immunosorbent reaction. Occasionally, the results may be falsified, which is most often caused by inflammation of the bladder. It is worth noting that research on the NMP-22 protein is still ongoing, which shows that despite the more and more common use of it as a marker of neoplastic changes in the bladder, numerous auxiliary tests and cystoscopy, which is currently the most reliable, are still performed when a positive result is obtained. a way of determining neoplastic changes in the bladder. However, work on the NMP-22 protein can be used in the routine examination of the urinary bladder.

## References

Costantini M, Gallo G, Attolini G. Urinary Biomarkers in Bladder Cancer. *Methods Mol. Biol.* 2021;2292:121-131.

Li S, Yue S, Y u C, Chen Y, Yuan D, Yu Q. A label-free immunosensor for the detection of nuclear matrix protein-22 based on a chrysanthemum-like Co-MOFs/CuAu NWs nanocomposite. *Analyst.* 2019 Jan 21;144(2):649-655.

Mati Q, Qamar S, Ashraf S, Khokhar M A, Arshad U. Tissue Nuclear Matrix Protein Expression 22 in Various Grades and Stages of Bladder Cancer. *J. Coll. Physicians Surg. Pak.* 2020 Dec;30(12):1321-1325.

Miyake M, Morizawa Y, Hori S, Tatsumi Y, Onishi S, Owari T, Iida K, Onishi K, Gotoh D, Nakai Y, Anai S, Chihara Y, Torimoto K, Aoki K, Tanaka N, Shimada K, Konishi N, Fujimoto K. Diagnostic and prognostic role of urinary collagens in primary human bladder cancer. *Cancer Sci.* 2017 Nov;108(11):2221-2228.

Sajid M T, Zafar M R, Ahmad H, Ullah S, Mirza Z I, Shahzad K. Diagnostic accuracy of NMP 22 and urine cytology for detection of transitional cell carcinoma urinary bladder taking cystoscopy as gold standard. *Pak. J. Med. Sci.* 2020 May-Jun;36(4):705-710.

# Chapter 2

# Carbohydrate Antigen 15.3 (CA 15.3)

**Natalia Żyłka**
**David Aebisher**
**and Dorota Bartusik-Aebisher***
Medical College of the University of Rzeszów, Poland

## Abstract

Carbohydrate antigen 15.3 or cancer antigen 15.3, (CA 15.3) is one of the most commonly used tumor markers. As many studies have shown, CA 15.3 is an invaluable biological marker. It is widely used in oncological diagnostics, but also in pulmonology. The 15.3 antigen is used to detect neoplastic diseases (breast cancer, ovarian cancer or lung cancer) and lung fibrosis, systemic sclerosis and rheumatoid arthritis. Disease prevention is as important as diagnosis. Contrary to appearances, the most important to perform regular preventive examinations adapted to the age and health condition of the patient (annual blood tests, and in women: cytology, ultrasound of breasts and reproductive organs). Genetic testing may be helpful if there has been a previous history of cancer in the patient's family.

**Keywords:** carbohydrate antigen 15.3 (CA 15.3), tumor marker, bladder, colon cancer, lung cancer, ASCO (American Society of Clinical Oncology)

* Corresponding Author's Email: dbartusikaebisher@ur.edu.pl.

In: The Medical Biology Guide to Proteins
Editor: David Aebisher
ISBN: 979-8-88697-910-7

Carbohydrate antigen 15.3 or cancer antigen 15.3, (CA 15.3) is one of the most commonly used tumor markers. It is a glycoprotein product of the MUC-1 gene and is present in the cell membrane of neoplastic epithelial cells. CA 15.3 is widely used in the diagnosis of breast cancer; however, it does not show organ specificity, which is characteristic of most marker proteins. As a consequence, its level may also be elevated in the course of other types of cancer, for example ovarian cancer, endometrial cancer, tubal cancer, pancreatic cancer, colon cancer, lung cancer, and even other non-cancerous diseases. CA 15.3 is a plasma mucin marker, but due to its low sensitivity in the early stages of malignancy, its presence and increased plasma concentration usually do not offer good diagnostic prospects. A consequence of organ non-specificity is that CA 15.3 is not suitable for screening but is instead widely used for the detection of recurrence and distant metastasis. For this reason, it can be used for postoperative follow-up of asymptomatic patients with malignant breast cancer. CA 15.3 is especially valuable in monitoring the treatment of patients with cancer that cannot be assessed diagnostically by existing radiological procedures (Lin et al. 2018).

CA 15.3 is a polymorphic mucin that belongs to the glycoproteins composed of a protein core (apomucin) and hydrocarbon chains. Nine genes encoding epithelial mucins are known: MUC1-MUC4, MUC5A / C, MUCC5B and MUC6-MUC8.

The most popular mucin most often described in studies and in the literature it is MUC1 (also known as Epsian mucin). It is encoded by the MUC1 gene on the first chromosome (1q21). The product of transcription and subsequent translation of the MUC1 gene is, inter alia, the CA 15.3 protein, resulting from non-homologous recombination of MUC1 fragments. This gene has tandem VNTR (variable-like associated antigen) repeats in its locus.

In the urinary bladder, pancreas and breast glands, mucins show a laminate arrangement in epithelial cells. During neoplastic transformation, the structure of the epithelium is damaged, which causes its abnormal polarity and enters the bloodstream (Magalhães et al. 2021).

Research led by Ebeling, who analyzed 1,046 surgically treated patients, made a significant contribution to the biology of mucins.

due to breast cancer. By carrying out concentration measurements before mastectomy or lumpectomy and during further observations (the median follow-up time was about three years), it was found that in a single-factor analysis, a high pre-operative CA 15.3 level may predict faster relapse and patient death as a result of the neoplastic process. On the other hand, the multivariate analysis, which additionally included the assessment of tumor size, periocular lymph nodes, histological grade G and estrogen receptor activity, did not show such utility. Despite the fact that a large number of scientific studies have proven the increased level of CA 15-3 in the course of breast cancer, the usefulness of determining the concentration of this marker in patients remains a controversial issue. Many tests using monoclonal antibodies have been developed that allow the detection of epitopes of the MUC1 gene product in the peripheral blood, but due to the low sensitivity and specificity of the above marker, it is possible to obtain a false result. As reported by ASCO (American Society of Clinical Oncology), the currently available literature does not allow CA 15.3 to be qualified as an optimal biochemical marker to be used in the screening and diagnosis of breast cancer (Moll et al. 2020).

The tumor marker CA 15.3 is used in patients with breast cancer to monitor the progress of hormone therapy and chemotherapy, to prevent people at genetic risk of developing breast cancer, and to control patients in order to detect local recurrence or distant metastases. Increased concentration of CA 15.3 in the peripheral blood is observed in approximately 10% of women with non-metastatic breast cancer and in approximately 70% of patients with distant metastases. Slightly elevated levels of this marker can also be noted in healthy people and in people with neoplasms other than breast cancer, e.g., colorectal cancer, lung cancer, as well as in non-neoplastic diseases (hepatitis, cirrhosis). The increasing concentration of the marker may indicate a lack of therapeutic effect of the selected antitumor treatment or a relapse of the disease. CA 15.3 monitoring should be started prior to treatment. especially when the advanced stage of the disease is suspected (Moll et al. 2020).

CA 15.3 is associated with fibroblastic activity. Its role as a biomarker has been confirmed in the diagnosis of pulmonary fibrosis, systemic sclerosis and rheumatoid arthritis. As fibroblastic activity reflects the secretion of CA 15.3 by type II pneumocytes, it is possible that the level of CA 15.3 can be used to determine the efficacy of drugs used in these lung diseases (Zajkowska et al. 2020).

As many studies have shown, CA 15.3 is an invaluable biological marker. It is widely used in oncological diagnostics, but also in pulmonology. As mentioned above, the 15.3 antigen is used, inter alia, to detect neoplastic diseases (breast cancer, ovarian cancer or lung cancer) and lung fibrosis, systemic sclerosis and rheumatoid arthritis. It is worth emphasizing, however, that CA 15.3 is used to diagnose existing diseases or their metastases or recurrences, which makes it impossible to use this marker in prophylaxis. Disease prevention is as important as diagnosis. So, what can be done to reduce the likelihood of cancer? Contrary to appearances, the most important are the behaviors and habits that people can implement themselves. The key is to lead a broadly understood healthy lifestyle. A properly balanced diet, daily physical effort of at least 30 minutes, avoiding stimulants such as alcohol or cigarettes can significantly increase the chances of avoiding many diseases, including cancer. It is also very important to perform regular preventive examinations adapted to the age and health condition of the patient. (Annual blood tests, and in women: cytology, ultrasound of breasts and reproductive organs). Genetic testing may be helpful if there has been a previous history of cancer in the patient's family.

In conclusion, testing the concentration of CA 15.3 in patients turned out to be a very useful tool in various fields of medicine, as confirmed by numerous studies. However, one should not forget about prophylaxis, especially cancer prevention.

## References

Lin DC, Genzen JR. Concordance analysis of paired cancer antigen (CA) 15-3 and 27.29 testing. *Breast Cancer Res Treat*. 2018 Jan;167(1):269-276.

Magalhães JS, Jammal MP, Crispim PCA, Murta EFC, Nomelini RS. Role of biomarkers CA-125, CA-15.3 and CA-19.9 in the distinction between endometriomas and ovarian neoplasms. *Biomarkers*. 2021 May;26(3):268-274.

Moll SA, Wiertz IA, Vorselaars AD, Zanen P, Ruven HJ, van Moorsel CH, Grutters JC. Serum biomarker CA 15-3 as predictor of response to antifibrotic treatment and survival in idiopathic pulmonary fibrosis. *Biomark Med*. 2020 Jul;14(11):997-1007.

Moll SA, Wiertz IA, Vorselaars ADM, Ruven HJT, van Moorsel CHM, Grutters JC. Change in Serum Biomarker CA 15-3 as an Early Predictor of Response to Treatment and Survival in Hypersensitivity Pneumonitis. *Lung*. 2020 Apr;198(2):385-393.

Zajkowska M, Gacuta E, Lubowicka E, Szmitkowski M, Ławicki S. Can VEGFR-3 be a better tumor marker for breast cancer than CA 15-3? *Acta Biochim Pol*. 2020 Mar 11;67(1):25-29.

# Chapter 3

# Thyroglobulin

**Izabela Niziołek**
**David Aebisher**
**and Dorota Bartusik-Aebisher***
Medical College of the University of Rzeszów, Rzeszów, Poland

## Abstract

Thyroglobulin is a substance classified as relatively large glycoprotein with a mass of 660 kDa. Functionally, it is a protein hormone synthesized by the epithelial cells of the thyroid gland, which, further secreted into the lumen of the follicles of this organ. It is estimated that up to 5% of the total number of people on Earth suffer from diseases related to the malfunctioning of the thyroid gland. This gland is especially important for the proper development of the body. The secretion products of this gland affect the operation of almost all systems of the human body. After oncological treatment and total thyroidectomy, the concentration of these antibodies should decrease, but if they are not, and their level increases, it indicates an increased risk of persistent cancer or its recurrence. Further research on this fundamental molecule of the thyroid gland may lead to the development of innovative methods enabling a more precise assessment of the condition of many patients, especially those struggling with thyroid abnormalities or its neoplasms.

**Keywords:** thyroglobulin, iodoglycoprotein, carboxylesterases, glycosylation, immunological test

* Corresponding Author's Email: dbartusikaebisher@ur.edu.pl.

In: The Medical Biology Guide to Proteins
Editor: David Aebisher
ISBN: 979-8-88697-910-7

Thyroglobulin is a substance classified as a relatively large glycoprotein with a mass of 660 kDa. Functionally, it is a protein hormone synthesized by the epithelial cells of the thyroid gland, which, further secreted into the lumen of the follicles of this organ, constitutes the most numerous (even up to 80% of all proteins present in the thyroid gland) group of proteins in this gland. Thyroglobulin (Thyroglobulin) is abbreviated as TG. This glycoprotein is a particularly important factor determining the proper functioning of one of the most important endocrine glands, i.e., the thyroid gland, because as a hormonal precursor it is responsible for the synthesis and storage of active hormones in this organ. Active thyroid hormones, T3-triiodothyronine and T4-thyroxine are responsible for the development, growth and maintenance of metabolism at the correct level, which makes TG, as their precursor, such an important component of the human body. The normal concentration of thyroglobulin in the blood in a healthy person is 1–30 μg / l. When blood levels of TG are elevated, it may indicate a significant degree of destruction of the follicles in which it is stored. Therefore, the determination of the level of this hormone in the blood is used as part of the laboratory diagnosis for inflammation or thyroid neoplasm, for which the level of TG is one of the most significant markers (Algeciras-Schimnich et al., 2018).

Thyroglobulin is, more precisely, an iodoglycoprotein because, in addition to carbohydrates accounting for 8% to 10%, it contains 0.1% to 2% iodine. It exists in the form of a soluble dimer. About 60 disulfide bonds per monomer and 17 glycosylation sites in the molecule are responsible for the solubility and extraordinary stability of this glycoprotein. Tg is enriched with tyrosine residues and is both a precursor to the main thyroid hormones and a storage facility for them. It belongs to the type B carboxylesterases - the lipase family. Tg biosynthesis occurs in the endoplasmic reticulum, and is regulated by TSH, insulin, and insulin-like growth factor 1 (IGF-1). Additionally, Tg synthesis undergoes autoregulation by reducing, under the given conditions, the expression of specific genes due to the inhibition of the transcription factor. Post-translational processing is of particular importance for thyroglobulin, as it results in micro molecular differences in Tg. Glycosylation, phosphorylation and iodination are responsible for this heterogeneity. These processes are important because they enable it to continue functioning, transporting and reacting. Thyroglobulin has 4 major hormonal sites on each of the constituent monomers of the final dimer. They are marked with letters A-D. The serum Tg half-life was determined to be 1 to 3 days (Bílek et al., 2020).

The concentration of Tg in the serum depends mainly on 3 factors, these are the differentiation presented by the thyroid tissue, the range of stimulation of the thyrethropic receptors, and any damage or inflammatory processes in the gland. In the case of a deficiency of the element iodine, in diseases characteristic of this condition, e.g., thyroid hyperplasia, there is also an increase in thyroglobulin concentration, similarly to its inflammation, or neoplasms, where the thyroid follicles are damaged and destroyed. The correlations found between the Tg level and iodine supply, the Tg level and the state of pregnancy or the upper limit of the normal Tg value as well as age and gender indicate that the use of thyroglobulin determination has a chance for a much wider use as an indicator of patients' condition than commonly, i.e., mainly to determine the state of patient with neoplastic disease of the thyroid gland. In deviations related to abnormal levels of thyroglobulin in the blood, the aim is to stabilize the functioning of the thyroid gland, which in turn affects the normalization of the Tg level (Citterio et al., 2019).

Unfortunately, problems arise in connection with the determination of Tg immunoanalytically. In current practice, an immunological test is used. As there is a noticeable diversity of dimer molecules in its isoforms, which is related to differences in the original structure as well as carbohydrate and iodine content, this test is one of the tests that is difficult to carry out and analyze. The above-mentioned factors that differentiate individual molecules from one another are responsible for the determination of the three-dimensional form, and therefore they can affect the reduction of antigenic determinants (epitopes) important in immunological analysis. Moreover, the presence of autoantibodies to Tg may have a limited impact on the Tg test. The limiting effect on the immunological test is understood as a reduction in the accuracy of the test, which consists of the precision of the determination and the accuracy of the obtained results of the analysis. Degrading the accuracy may result in falsely low results. There is another test that is more resistant to Tg autoantibodies, but with this test the result may give false high data. Interestingly, the level of autoantibodies is influenced by the intake of iodine in the body. Increasing its consumption stimulates an increase in the amount of autoantibodies. A promising method for determining Tg is mass spectrometry, in which autoantibodies would not interfere with the results of the analyzes (Coscia et al., 2020). Numerous studies and analyzes have been carried out, but they did not give a definite answer as to how thyroglobulin is recognized by the immune system. It has been proven that the immune response is relatively limited, however, the recognition of pathogenic fragments by autoantibodies, as well as their localization in the Tg molecule, is still not sufficiently studied and requires further research (Tourani et al., 2021).

It is estimated that up to 5% of the total number of people on Earth suffer from diseases related to the malfunctioning of the thyroid gland. This gland is especially important for the proper development of the body. The thyroid gland is formed very early, because already on the 22nd day of embryogenesis, it comes from the endoderm and is the first endocrine structure. The secretion products of this gland affect the operation of almost all systems of the human body.

Antibodies against Tg are the biggest problem for laboratory diagnosticians because they falsely lower Tg in the serum tested by immunometric tests. These antibodies are found in about a quarter of people with thyroid cancer, but also in about 10% of the general population. After oncological treatment and total thyroidectomy, the concentration of these antibodies should decrease, but if they are not, and their level increases, it indicates an increased risk of persistent cancer or its recurrence. All the information presented indicates the importance of the presence of thyroglobulin in the body, its enormous impact on its proper development, as well as the marker importance of the detection of serious disorders and diseases. Further research on this fundamental molecule of the thyroid gland may lead to the development of innovative methods enabling a more precise assessment of the condition of many patients, especially those struggling with thyroid abnormalities or its neoplasms.

## References

Algeciras-Schimnich A. Thyroglobulin measurement in the management of patients with differentiated thyroid cancer. *Crit. Rev. Clin. Lab. Sci*. 2018;55(3):205-218.

Bílek R, Dvořáková M, Grimmichová T, Jiskra J. Iodine, thyroglobulin and thyroid gland. *Physiol. Res*. 2020;69(Suppl 2): S225-S236.

Citterio C E, Targovnik H M, Arvan P. The role of thyroglobulin in thyroid hormonogenesis. *Nat. Rev. Endocrinol*. 2019;15(6):323-338.

Coscia F, Taler-Verčič A, Chang V T, Sinn L, O'Reilly F J, Izoré T, Renko M, Berger I, Rappsilber J, Turk D, Löwe J. The structure of human thyroglobulin. *Nature*. 2020;578(7796):627-630.

Tourani S S, Fleming B, Gundara J. Value of thyroglobulin post hemithyroidectomy for cancer: a literature review. *ANZ J. Surg*. 2021;91(4):724-729.

# Chapter 4

# Carcinoembryonic Antigen (CEA)

**Paweł Woźnicki**
**Dorota Bartusik-Aebisher***
**and David Aebisher**
Medical College of the University of Rzeszów, Poland

## Abstract

Carcinoembryonic antigen (CEA) is one of the most used tumor markers in the world. CEA is a group of foetal glycoproteins produced in the developing tissues of the foetal digestive tract, pancreas and liver. Their synthesis is almost completely stopped before birth and in healthy adults they are not produced in significant amounts, therefore the level of CEA in the blood is very low, less than 5ng / ml. CEA can be designated as a non-specific cancer marker. The discovery of this antigen initiated a series of studies enabling the use of the characteristics of these glycoproteins in the diagnosis and monitoring of the course of malignant colon cancer.

**Keywords:** carcinoembryonic antigen (CEA), cancer-fetal antigen, glycoproteins, signal transduction

Carcinoembryonic antigen (CEA) is one of the most used tumor markers in the world. CEA was first isolated from colon cancer tissue and described in 1965 by Phil Gold and Samuel O. Freedman. CEA is a group of fetal

* Corresponding Author's Email: dbartusikaebisher@ur.edu.pl.

In: The Medical Biology Guide to Proteins
Editor: David Aebisher
ISBN: 979-8-88697-910-7

glycoproteins produced in the developing tissues of the fetal digestive tract, pancreas and liver. Their synthesis is almost completely stopped before birth and in healthy adults they are not produced in significant amounts, therefore the level of CEA in the blood is very low, less than 5ng / ml. CEAs are encoded by 29 genes located on the long arm of chromosome 19, but only 18 of them are expressed (Campos-da-Paz et al. 2018). The carcinoembryonic antigen is bound to the cell membrane and shows a complex pattern of expression in normal and neoplastic tissues. These glycoproteins are involved in cellular adhesion as well as possibly in signal transduction or regulation. Neoplastic cells of the intestine, pancreas, liver, stomach, and in some cases also the lungs, produce carcino-fetal antigen in an increased amount. Then it increases its level in the blood. Therefore, CEA can be designated as a non-specific cancer marker. An increase in the level of carcinoembryonic antigen in the blood also occurs with smoking, pancreatitis, biliary obstruction, peptic ulcer and hypothyroidism, but the degree of increase is much less than in cancer (de Melo et al. 2020).

They managed to show that the group of these antigens was present in the changed neoplastic cells as well as in the fetal intestine, fetal liver and pancreas between 2 and 6 months of pregnancy, while it was not present in healthy adult tissues. Hence, a group of these glycoproteins was called the carcinoembryonic antigen (Konishi et al. 2018).

CEA is a group of glycoproteins with an approximate mass of 180 kDa. The carcinoembryonic antigen is bound to the cell membrane. It shows a complex pattern of expression in healthy and pathological tissues. The role of CEA is to participate in cellular adhesion as well as possibly in signal transduction or regulation. CEAs are believed to be involved in the mechanism of innate immunity, protecting the large intestine, as well as possibly the upper gastrointestinal tract and the bladder from microbial attack. These glycoproteins are encoded by 29 genes, which are located on the long arm of chromosome 19. Only 18 of them are expressed - 7 belonging to the CEA subgroup and 11 belonging to the pregnancy-specific glycoprotein subgroup. They are produced in the embryonic tissues of the digestive tract, pancreas and liver. Their synthesis is almost completely stopped before birth and in healthy adults they are not produced in significant amounts. Since 1965, nearly 23,000 publications have been published. research papers on the carcinoembryonic antigen. The number of articles published showed an upward trend until 1990. Subsequently, the number of items issued decreased. Since the beginning of the 21st century, we have seen a steady increase in the number of publications on CEA (Li et al. 2020).

Cancer-fetal antigen is one of the most widely used tumor markers in the world. Elevated levels of CEA in the blood are found in certain malignancies: colon, stomach, pancreas, lung, breast and medullary thyroid cancer, certain diseases such as inflammatory bowel disease (ulcerative colitis, Crohn's disease), pancreatitis and cirrhosis liver as well as in smokers. Therefore, testing the level of CEA in the blood is not reliable as an early detection of cancer. Carcino-fetal antigen measurement is used in modern diagnostics as a marker of the treatment of malignant colorectal neoplasms and to determine the stage of cancer spread. After successful surgical removal of the tumor, the level of CEA in the blood should return to normal (below 5 ng/ml), therefore CEA testing is used to monitor the effectiveness of the procedure.

In tumors with high levels of carcinoembryonic antigen expression, anti-CEA antibodies can be used for targeted anti-cancer therapy (Mikkelsen et al. 2019).

The discovery of this antigen initiated a series of studies enabling the use of the characteristics of these glycoproteins in the diagnosis and monitoring of the course of malignant colon cancer. A study that accurately describes the impact of CEA tests on overall survival, cost of treatment and its effects has never been conducted. However, based on the available results, it can be concluded that the monitoring of all colorectal cancer patients with serial CEA tests has little effect on treatment outcomes. The question that therefore needs to be asked is whether the detection of carcinoembryonic antigen in the blood should occur in all cases, or only as a marker of treatment and for the determination of the severity of the spread of cancer.

Anti-CEA antibodies have also found application in targeted anti-cancer therapy in the treatment of colorectal cancer.

## References

Campos-da-Paz M, Dórea JG, Galdino AS, Lacava ZGM, de Fatima Menezes Almeida Santos M. Carcinoembryonic Antigen (CEA) and Hepatic Metastasis in Colorectal Cancer: Update on Biomarker for Clinical and Biotechnological Approaches. *Recent Pat Biotechnol*. 2018;12(4):269-279.

Izabella Abreu de Melo M, Rodrigues Correa C, da Silva Cunha P, Miranda de Góes A, Assis Gomes D, Silva Ribeiro de Andrade A. DNA aptamers selection for carcinoembryonic antigen (CEA). *Bioorg Med Chem Lett*. 2020 Aug 1;30(15): 127278.

Konishi T, Shimada Y, Hsu M, Tufts L, Jimenez-Rodriguez R, Cercek A, Yaeger R, Saltz L, Smith JJ, Nash GM, Guillem JG, Paty PB, Garcia-Aguilar J, Gonen M, Weiser MR.

Association of Preoperative and Postoperative Serum Carcinoembryonic Antigen and Colon Cancer Outcome. *JAMA Oncol.* 2018 Mar 1;4(3):309-315.

Li W, Ma C, Song Y, Hong C, Qiao X, Yin B. Sensitive detection of carcinoembryonic antigen (CEA) by a sandwich-type electrochemical immunosensor using MOF-Ce@HA/Ag-HRP-Ab<sub>2</sub> as a nanoprobe. *Nanotechnology.* 2020 May 1;31(18):185605.

Mikkelsen K, Harwood SL, Compte M, Merino N, Mølgaard K, Lykkemark S, Alvarez-Mendez A, Blanco FJ, Álvarez-Vallina L. Carcinoembryonic Antigen (CEA)-Specific 4-1BB-Costimulation Induced by CEA-Targeted 4-1BB-Agonistic Trimerbodies. *Front Immunol.* 2019 Jul 31;10:1791.

# Chapter 5

# Carbohydrate Antigen 125 (CA 125)

**Agnieszka Przygórzewska**
**Dorota Bartusik-Aebisher***
**and David Aebisher**
Medical College of the University of Rzeszów, Poland

## Abstract

Carbohydrate antigen 125 (CA-125), also known as MUC16, is a high molecular weight soluble glycoprotein belonging to the mucin family (MUC). CA-125 was first discovered in an ovarian cancer cell line. There are also hypotheses that it is also secreted by tissues of neoplastic origin. CA-125 has long been used as a marker for ovarian tumors. Elevated serum levels of CA-125 may also indicate other neoplasms such as lung and mediastinal cancer, teratoma or non-Hodgkin's lymphoma. Monitoring the concentration of carbohydrate antigen 125 in the blood makes it possible to predict the likelihood of this cancer developing. Long-term control of CA125 concentration enables segregation of patients into groups with low, medium and high risk of ovarian cancer development, which is of great importance in the early detection of the disease. Due to the high frequency of this disease, it is important to further develop research on the carbohydrate antigen 125.

**Keywords:** carbohydrate antigen 125 (CA-125), lung and mediastinal cancer, teratoma, non-Hodgkin's lymphoma

---

* Corresponding Author's Email: dbartusikaebisher@ur.edu.pl.

In: The Medical Biology Guide to Proteins
Editor: David Aebisher
ISBN: 979-8-88697-910-7

Carbohydrate antigen 125 (CA-125), also known as MUC16, is a high molecular weight soluble glycoprotein belonging to the mucin family (MUC). Human CA-125, encoded by the MUC16 gene, contains approximately 22,000 amino acids and is highly glycosylated in the extracellular region. CA-125 was first discovered in an ovarian cancer cell line. It is normally expressed on the surface of cells derived from celomic epithelium, including the pleura, epicardium, fallopian tubes, endometrium, and cervix. There are also hypotheses that it is also secreted by tissues of neoplastic origin (Bonifácio et. al. 2020). The physiological role of CA-125 is believed to protect epithelial surfaces from physical stress through a moisturizing process. CA-125 has long been used as a marker for ovarian tumors. Determination of its serum concentration is helpful in monitoring this neoplasm and further stratification of risk and prognosis after the end of treatment. Elevated serum levels of CA-125 may also indicate other neoplasms such as lung and mediastinal cancer, teratoma or non-Hodgkin's lymphoma. In addition to cancer, elevation of CA-125 can be observed in several other physiological or pathological conditions, including early pregnancy, menstruation, and peritoneal and ascites injuries. Carbohydrate antigen 125 also plays a significant role in heart diseases, and its increased level can be observed, among others, in patients with chronic heart failure (Bonifácio et. al. 2020). CA125 is composed of a short cytoplasmic tail, a transmembrane domain and a large extracellular structure with extensive glycosylation. This antigen was initially detected using the murine monoclonal antibody designated OC125. Robert Bast, Robert Knapp and his research team first isolated this antibody in 1981. The protein was dubbed the "cancer antigen 125" because OC125 was an antibody produced against the ovarian cancer cell line under study (Dochez et. al. 2019).

An increase in CA125 concentration (above the cut-off value of 30–35 U / ml) is observed in serially drawn blood samples from women with ovarian cancer. So for screening, long-term monitoring of CA125 levels is likely to be more useful than a single measurement. The ovarian cancer risk algorithm predicts the probability of ovarian cancer on the basis of long-term monitoring of CA125, allowing segregation of patients into groups with low, medium and high risk of ovarian cancer development. Increased CA125 concentration may also suggest recurrence of the disease. The CA125 test is also used to distinguish between benign pelvic tumors and ovarian cancer. All indications are that CA125 is and will remain the main and most clinically significant biomarker of ovarian cancer in the near future (Dochez et. al. 2019).

Correct interpretation of elevated CA-125 levels in heart failure should take into account both triggers and the site of production. These issues have been investigated and no simple answer has been found. Until now, it has been assumed that CA-125 production in individuals with HF occurs with so-called "stressed" mesothelial cells in response to both haemodynamic and inflammatory stimuli. It has been suggested that fluid overload, accompanied by high venous pressure due to heart failure, may increase congestion and hydrostatic pressure in the mesothelium, which may result in the release of inflammatory markers such as interleukin-6 (IL-6), interleukin-10 (IL-10) and tumor necrosis factors (Zhang et. al. 2021). Thus, irritation of the serous membranes due to inflammation, mechanical stress, or fluid congestion may stimulate mesothelial cells to initiate carbohydrate antigen synthesis 125 (Falcão et. al. 2020). The pathological processes may ultimately contribute to the activation of the neurohumoral axis, facilitating the production of CA-125. The displacement of bacteria or the formation of endotoxins sometimes occurring during exacerbation of heart failure may also play an important role (Santas et. al. 2020). There was also a hypothetical relationship between pericardial stimuli and increases in carbohydrate antigen 125 levels in people with heart failure. The CA-125 glycoprotein, previously known mainly as a diagnostic marker for ovarian cancer screening and therapeutic monitoring, is gaining more and more attention due to its usefulness in diagnosing other diseases, including heart failure (Santas et. al. 2020).

Ovarian cancer accounts for about 4.6% of cancer cases among women in Poland. The mortality rate from this disease is approximately 6% of all cancer deaths in women. Late diagnosis and hence poor prognosis are a problem in the successful treatment of ovarian cancer. More than half of ovarian malignant neoplasms occur in women aged 50–70 years, and approximately 15% of cases occur in women under 50 years of age. Monitoring the concentration of carbohydrate antigen 125 in the blood makes it possible to predict the likelihood of this cancer developing. Long-term control of CA125 concentration enables segregation of patients into groups with low, medium and high risk of ovarian cancer development, which is of great importance in the early detection of the disease. CA125 can also distinguish between ovarian cancer and pelvic tumors. Therefore, it has a significant impact on the diagnosis and monitoring of the development of this cancer.

This disease affects about 1-2% of the population, and its incidence rapidly increases after the age of 75, and in people aged 70-80 it is even up to 20%. Based on epidemiological data, it can be assumed that around 600,000 people suffer from heart failure in Poland. Up to 1 million people. Due to the high frequency of this disease, it is important to further develop research on the carbohydrate antigen 125.

## References

Bonifácio VDB. Ovarian Cancer Biomarkers: Moving Forward in Early Detection. *Adv Exp Med Biol*. 2020; 1219:355-363.

Dochez V, Caillon H, Vaucel E, Dimet J, Winer N, Ducarme G. Biomarkers and algorithms for diagnosis of ovarian cancer: CA125, HE4, RMI and ROMA, a review. *J Ovarian Res*. 2019 Mar 27;12(1):28.

Falcão F, Oliveira F, Cantarelli F, Cantarelli R, Brito Júnior P, Lemos H, Silva P, Camboim I, Freire MC, Carvalho O, Sobral Filho DC. Carbohydrate antigen 125 for mortality risk prediction following acute myocardial infarction. *Sci Rep*. 2020 Jul 3;10(1):11016.

Santas E, Palau P, Bayés-Ge A, Núñez J. The emerging role of carbohydrate antigen 125 in heart failure. *Biomark Med*. 2020 Mar;14(4):249-252.

Zhang M, Cheng S, Jin Y, Zhao Y, Wang Y. Roles of CA125 in diagnosis, prediction, and oncogenesis of ovarian cancer. *Biochim Biophys Acta Rev Cancer*. 2021 Apr;1875(2):188503.

# Chapter 6

# Calcitonin

**Wiktor Słaby**
**Dorota Bartusik-Aebisher***
**and David Aebisher**
Medical College of the University of Rzeszów, Poland

## Abstract

Calcitonin (CT) is an animal polypeptide hormone that consists of 32 amino acids. The mass of a single molecule is 3.42 kDa. The basic role of the hormone is to regulate the phosphorus and calcium balance of the body. For a long time in the history of calcitonin research, the role of this hormone has given rise to scientific debate about its exact action and effect on living organisms. This was mainly due to the fact that its deficiency or excess was not observed in patients with diseases of the skeletal system, although its physiological effects were documented. The exact role it plays in human physiology is not yet fully understood. In laboratory studies, the most important reason for determining its level is still the study of medullary thyroid cancer (MCT), a disease for which calcitonin is a sensitive and specific marker.

**Keywords:** calcitonin (CT), hormone, peribalveal C cells, radioimmunoassay method, Hypercalcemia, calcitonin receptors (CTRs), medullary thyroid cancer (MCT)

---

* Corresponding Author's Email: dbartusikaebisher@ur.edu.pl.

In: The Medical Biology Guide to Proteins
Editor: David Aebisher
ISBN: 979-8-88697-910-7

Calcitonin (CT) is an animal polypeptide hormone that consists of 32 amino acids. The mass of a single molecule is 3.42 kDa. The basic role of the hormone is to regulate the phosphorus and calcium balance of the body. It is produced and secreted mainly by C cells of the thyroid gland in response to the increase in calcium concentration in the extracellular space. It can also form to a lesser extent in the nervous system, in the lungs and in the digestive system. It is formed from the transformation of its prohormone, i.e., procalcitonin. Its main action is to reduce bone resorption by osteoclasts, i.e., transformed monocytes found in bone tissue. Calcitonin reduces the rate of calcium absorption and transport by osteoclasts, and also slows down the release of calcium ions into the intercellular fluid. Measurements of circulating calcitonin using a radioimmunoassay method indicate that calcitonin is not secreted until plasma calcium levels are approximately 9.5 mg / dl. Moreover, calcitonin stimulates the intestinal reabsorption of calcium and the production of 1,25 (OH) 2D, i.e., the provitamin vitamin D3 (Allison et al., 2019).

All these actions lower the calcium level in the body. The blood test for calcitonin is not a routine diagnostic test, but measuring calcitonin plays a role in between others in the diagnosis of thyroid neoplasms. For example, in the case of medullary thyroid cancer, the level of calcitonin in the patient's blood may be significantly increased. Hypercalcemia, i.e., elevated levels of calcium in the blood, is the basic stimulus that stimulates the release of calcitonin. Calcitonin was discovered in 1962 by two researchers, Harold Copp and B. Cheney. In 1964 A. Baghdiantz and his colleagues developed a method for the isolation of calcitonin. Pure calcitonin was obtained from pig thyroid in six laboratories in Switzerland and the United States, and the chemical structure of the purified hormone was established in three of these laboratories (Allison et al., 2019).

Research into the secondary and tertiary structure of porcine calcitonin was carried out in 1970 by researchers Brewer and Edelhoch, who reported that in the aquatic environment the hormone is approximately 90 percent disordered and the remainder in an ordered alpha-helix configuration. As for the structure of calcitonin in non-polar solvents, such as 2-chloroethanol, as much as 50 percent of the polypeptide is in the form of an ordered alpha-helix. This dependence may be related to the interaction of the hormone at receptor target sites, for example hydrophobic parts of cell membranes with a lipid structure. The most important role in the study of calcitonin in humans was its detection in the parenchyma of medullary thyroid carcinoma (MTC), which is a neuroendocrine malignant neoplasm originating in the peribalveal C cells of the thyroid gland. These cells come from the nerve crest, from where during fetal development they migrate from the fifth gill pocket to the thyroid gland, where they are responsible for the production of calcitonin. The high content of this hormone in the cells of the tumor parenchyma allowed for the isolation of human calcitonin and determination of its amino acid sequence (Hay et al., 2018).

The basic form - alpha calcitonin, consists of 32 amino acids, and its sequence is as follows: Cys-Gly-Asn-Leu-Ser-Thr-Cys-Met-Leu-Gly-Thr-Tyr-Thr-Gln-Asp-Phe-Asn-Lys-Phe-His-Thr-Phe-Pro-Gln-Thr-Ala-Leu-Gly-Val-Gly-Ala-Pro-$NH_2$. The position of the acidic and basic amino acids shows some regular regularity. The acidic ones are in positions 15 or 30, and the basic ones in positions 11, 14, 18 and 21. The molecule is linked by a disulfide bridge between the cysteine at the first position (Cys-1) and the cysteine at the 7 position (Cys-7). The structure of human calcitonin is most similar to that of salmon calcitonin, as the 16 amino acids in both polypeptides are identical. However, the differences mean that the intake of salmon calcitonin induces the production of antibodies in the majority of the patients to whom it was administered. Salmon calcitonin exceeds about 20 times the specific activity of human calcitonin. This activity is accompanied by a prolonged delay in blood calcium levels following administration of the hormone. In humans, there are four forms of this polypeptide, in addition to calcitonin alpha, we distinguish beta, gamma and delta calcitonin, which differ in the composition and number of amino acids (Srinivasan et al., 2020).

The secretion of calcitonin is physiologically regulated by the concentration of calcium ions on the basis of negative feedback. When the concentration of Ca2 + in the blood plasma increases, these ions directly affect the increased release of calcitonin, which is stored in the granules associated with the cell membrane of the thyroid C cells. Apart from calcium, cholecystokinin, gastrin, estrogens and glucagon have a stimulating effect on the release of this hormone. The main mechanism of action of calcitonin is inhibition of osteoclast activity in bone, resulting in increased deposition of calcium and phosphate in bone tissue. In addition, this hormone increases the activity of osteoblasts, which stimulates the synthesis of bone matrix and the binding of calcium ions in the bones due to an increase in the concentration of osteocalcin and osteonectin. Then the osteogenesis process becomes the dominant phenomenon in the bones. Calcitonin is an antagonist of parathyroid hormone, i.e., a hormone of the parathyroid gland that increases the concentration of calcium and phosphate ions in the blood plasma. In addition to its primary function, calcitonin has little effect on the urinary excretion of calcium, magnesium, and phosphate in the kidney, as well as on gastric and pancreatic juices. Its action on centers in the central nervous system suppresses appetite and has an analgesic effect by stimulating the release of endogenous opiates and increasing the number of their receptors in the hypothalamus and increasing the excitability threshold for pain stimuli (Naot et al., 2019).

The normal value of calcitonin in the blood of a healthy person is below 10 pg/ml. Determination of the concentration of calcitonin in the blood serum is an important test in patients with suspected medullary thyroid cancer (MCT), which is a specific marker of this malignant neoplasm. In over 90% of cases, the presence of MCT is associated with a significant increase in the concentration of this hormone. Medullary thyroid cancer is detected by immunohistochemistry using anti-calcitonin antibodies. Its determination enables preoperative cancer detection and is also a good tool for assessing the effectiveness of the treatment and monitoring the course of the disease. An increase in the concentration of calcitonin above 30 pg/ml of plasma may indicate the presence of neoplastic cells and is an indication for imaging tests. Different ranges are used in patients with medullary carcinoma. In MCT confined to the thyroid gland, calcitonin concentrations not exceeding 1000 pg / ml are observed. When the tumor is also present in the structures adjacent to the thyroid gland, the concentration of the hormone may be 1000-100000 pg / ml. In neoplastic disease with metastases to distant organs, the result of calcitonin concentration may exceed 100,000 pg / ml. Its level is also controlled in people with neuroendocrine tumors such as carcinoid tumors, insulinomas or small cell lung cancer. Calcitonin belongs to the so-called anti-resorptive drugs, i.e., drugs that reduce chaotic bone remodeling, slow down the pace of adverse changes and relieve bone pain. In the treatment of calcitonin deficiency, salmon calcitonin is used, due to its similarity to human calcitonin. More calcitonin is secreted in growing children, and this is probably related to its role in skeletal development. In addition, it protects the mother's bones against excessive loss of calcium during pregnancy. This is due to the building of the skeleton of the fetus and lactation, during which the loss of calcium by the woman's body is increased. The half-life of human calcitonin is approximately 10 minutes (Yazdani et al., 2019).

For a long time in the history of calcitonin research, the role of this hormone has given rise to scientific debate about its exact action and effect on living organisms. This was mainly due to the fact that its deficiency or excess was not observed in patients with diseases of the skeletal system, although its physiological effects were documented. A breakthrough in understanding this polypeptide came with the discovery of calcitonin receptors (CTRs), which are G-protein coupled and allow calcitonin to bind to osteoclasts. They are also found in kidney cells, where there is an increased excretion of calcium in the urine after attaching to calcitonin receptors. However, the determination of calcitonin is not used to assess disorders of calcium metabolism. The exact role it plays in human physiology is not yet fully understood. Its content in the human thyroid is relatively low, and after removal of the thyroid gland, the bone content and plasma calcium levels are usually normal, provided that the parathyroid glands and their parathyroid hormone function properly. The most important reason for determining its level is still the study of medullary thyroid cancer (MCT), a disease for which calcitonin is a sensitive and specific marker. There is general agreement that calcitonin has little effect after a longer duration of action on $Ca^{2+}$ concentrations in animals and humans.

## References

Allison SL, Davis KW 2nd. Calcitonin Stewardship Strategies. *J. Pharm. Pract.* 2019 Oct;32(5):584-585.

Hay DL, Garelja ML, Poyner DR, Walker CS. Update on the pharmacology of calcitonin/CGRP family of peptides: IUPHAR Review 25. *Br. J. Pharmacol.* 2018 Jan;175(1):3 17.

Naot D, Musson DS, Cornish J. The Activity of Peptides of the Calcitonin Family in Bone. *Physiol. Rev.* 2019 Jan 1;99(1):781-805.

Srinivasan A, Wong FK, Karponis D. Calcitonin: A useful old friend. *J. Musculoskelet. Neuronal Interact.* 2020 Dec 1;20(4):600-609. PMID: 33265089; PMCID: PMC77 16677.

Yazdani J, Khiavi RK, Ghavimi MA, Mortazavi A, Hagh EJ, Ahmadpour F. Calcitonina como agente analgésico: revisão dos mecanismos de ação e das aplicações clínicas [Calcitonin as an analgesic agent: review of mechanisms of action and clinical applications]. *Braz J. Anesthesiol.* 2019 Nov-Dec;69(6):594-604. Portuguese.

# Chapter 7

# Epidermal Growth Factor (EGF)

**Sara Śmiałek**
**Dorota Bartusik-Aebisher***
**and David Aebisher**
Medical College of the University of Rzeszów, Poland

## Abstract

Epidermal growth factor (EGF) is a small protein, composed of 53 amino acids and completely devoid of alanyl, phenylalanyl, or lysine residues. Its structure includes 3 internal disulfide bonds, and the hormone itself has a second-order structure, periodically assuming the form of a β-sheet. Extremely important protein needed to maintain homeostasis in the human body. It plays a fundamental role in stimulating epithelial cells to divide. Due to this property, intensive research has been carried out to determine the use of EGF in the wound healing process. The effect of using these antibodies is the increased sensitivity of cancer cells to radiotherapy, inhibition of tumor growth and induction of apoptosis processes in glioblastoma multiforme cells. Currently, laboratories around the world are working intensively to better understand the role of this factor in the human body, and to discover new, previously unknown properties that may significantly affect our lives in the future.

**Keywords:** epidermal growth factor (EGF), transforming growth factor α (TGFα), phosphotyrosines, receptor tyrosine kinase (RTK)

---

* Corresponding Author's Email: dbartusikaebisher@ur.edu.pl.

In: The Medical Biology Guide to Proteins
Editor: David Aebisher
ISBN: 979-8-88697-910-7

Epidermal growth factor (EGF) is a small protein, composed of 53 amino acids and completely devoid of alanyl, phenylalanyl, or lysine residues. Its structure includes 3 internal disulfide bonds, and the hormone itself has a second-order structure, periodically assuming the form of a β-sheet. It was first isolated by Dr. Stanley Cohen and Dr. Rita Levi-Montalcini from the submandibular glands of mice. Thanks to their discovery, they were awarded the Nobel Prize in 1986. They conducted a lot of research to understand the properties of EGF. This factor has been shown to be directly mitogenic and therefore independent of other hormones and factors. In turn, injecting newborn mice with raw preparations of the submandibular glands results in premature opening of the eyelids and eruption of teeth. It was concluded that EGF promotes cell division of epithelial tissue (Chen et al., 2018).

In 1973, they observed that murine EGF stimulated DNA synthesis in human fibroblasts, which could indicate the presence of a similar protein in our body. When human EGF was isolated, it was found to act in a similar way to its murine counterpart in promoting the proliferation of fibroblasts and corneal epithelial cells. Jerzy Kaulbersz and Stanisław J. Konturek found that urogastrone (a hormone that inhibits gastric secretion) is identical in structure to EGF. Both compounds elicit the same response in target cells, namely they strongly inhibit gastric acid secretion. This chapter describes the mechanism of EGF stimulation of target cell growth, its role in the human body and its clinical application (Fang et al., 2020).

EGF binds to a specific receptor - EGFR. This receptor, also called ErbB1 or HER1, is a transmembrane protein - type I receptor tyrosine kinase (RTK). ErB1 is a member of the ErbB receptor family (which in turn includes three other receptor tyrosine kinases: ErbB2 / Neu / HER2, ErbB3 / HER3, and ErbB4 / HER4). The epidermal growth factor receptor is activated when it binds to specific ligands. They can be, inter alia, the aforementioned EGF or transforming growth factor α (TGFα). Upon ligand binding, the receptor transforms from an inactive monomeric form to an active homodimer, or to an active heterodimer when it binds to another member of the ErB receptor family. Dimerization stimulates the autophosphorylation of several tyrosine residues in the receptor's C-terminal domain. This allows other signaling proteins to bind to the resulting phosphotyrosines via their own SH2 domains. There are two main intracellular pathways activated by EGFR which stimulate DNA synthesis and cell proliferation. The first, the RAS-RAF-MEK-MAPK pathway, controls the transition from the G1 phase to the S phase of the cell cycle. The second PI3K-Akt pathway activates the cascade of anti-apoptotic and cytoprotective mediators, thus protecting damaged cells. This pathway can be triggered by the exposure of keratinocytes to UV radiation (Kim et al., 2020).

Overexpression of EGFR is associated with the development of neoplastic diseases, including non-small cell lung cancer (40% of cases), rectal cancer, glioblastoma, which is the most aggressive type of brain cancer (50%), and epithelial tumors around the head and neck (80-100%). Specific EGFR inhibition is used in anti-cancer therapy. For this, various measures are used. These can be anti-EGFR monoclonal antibodies, e.g., cetuximab and panitumumab, which inhibit EGF binding, or small molecule receptor tyrosine kinase inhibitors that block EGFR activation and phosphorylation. Vaccines have also been developed to target EGF / EGFR dependent tumors. An example is the CimaVax-EGF vaccine, developed by the Center of Molecular Immunology in Cuba. It contains recombinant EGF coupled to a protein carrier enhancing the immune response. Introducing the vaccine into the patient's body stimulates the production of antibodies against EGF. This lowers the level of EGF in the blood, a factor necessary for the growth of cancer cells. Therefore, this vaccine is considered to be of greater importance when used prophylactically (Pudełek et al., 2020).

The epidermal growth factor EGF has also found use as a wound healing stimulant. Many studies suggest that EGF must be administered in a sufficiently high concentration and for a sufficiently long time to effectively heal a wound. EGF is upregulated in the event of acute wounds, which stimulates the expression of keratins (proteins produced by keratinocytes) and at the same time promotes epithelial reconstruction and tensile strength. This is not the case in chronic wounds where EGF is downregulated and EGFR shifts from the plasma membrane into the keratinocyte cytoplasm. This inhibits the process of epithelial reconstruction. This process is probably due to the easier degradation of EGF in the chronic wound environment, which limits the healing process. Therefore, during treatment, EGF is delivered in a more stable form, for example in the form of gels, hydrofibers, or in a form conjugated with other substances, such as hyaluronic acid, collagen or liposomes. The fact that EGF must be supplied in high concentrations raises concerns against the development of neoplastic lesions. However, studies to date have not shown any negative effects of EGF in wound healing (Richani et al., 2018).

Epidermal growth factor is a physiologically extremely important protein needed to maintain homeostasis in the human body. It plays a fundamental role in stimulating epithelial cells to divide. Due to this property, intensive research has been carried out to determine the use of EGF in the wound healing process. It is now used clinically to accelerate the healing of skin lesions.

The mechanisms of the cascades of reactions triggered in cells when EGF binds to its specific receptor have also been discovered. On this basis, it was found that in addition to its mitogenic effect, EGF also has the ability to protect damaged cells by inhibiting their apoptosis.

Also found in medicine was EGFR. The development of a vaccine against neoplasms dependent on the epidermal growth factor or its receptor has become an important undertaking.

Anti-EGFR monoclonal antibodies are gaining more and more application in medicine. They give hope in the treatment of cancers caused by overexpression of EGFR. Their special role in influencing stage IV neoplasm, such as glioblastoma, should be emphasized. The effect of using these antibodies is the increased sensitivity of cancer cells to radiotherapy, inhibition of tumor growth and induction of apoptosis processes in glioblastoma multiforme cells.

It is impossible to take into account the huge number of reports on EGF. Currently, laboratories around the world are working intensively to better understand the role of this factor in the human body, and to discover new, previously unknown properties that may significantly affect our lives in the future.

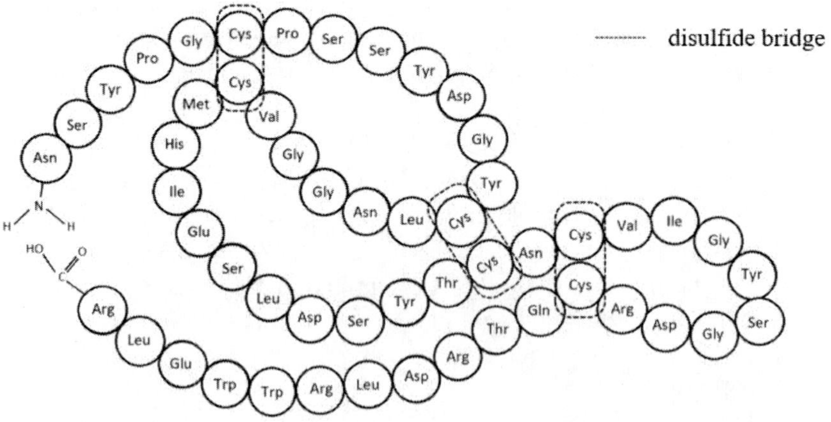


**Figure 1.** Amino acid sequence and epidermal growth factor disulfide bonds.

## References

Chen R, Jin G, Li W, McIntyre TM. Epidermal Growth Factor (EGF) Autocrine Activation of Human Platelets Promotes EGF Receptor-Dependent Oral Squamous Cell Carcinoma Invasion, Migration, and Epithelial Mesenchymal Transition. *J. Immunol.* 2018 Oct 1;201(7):2154-2164.

Fang T, Zhang Y, Chang V Y, Roos M, Termini CM, Signaevskaia L, Quarmyne M, Lin PK, Pang A, Kan J, Yan X, Javier A, Pohl K, Zhao L, Scott P, Himburg HA, Chute JP. Epidermal growth factor receptor-dependent DNA repair promotes murine and human hematopoietic regeneration. *Blood.* 2020 Jul 23;136(4):441-454.

Kim H Y, Um S H, Sung Y, Shim M K, Yang S, Park J, Kim ES, Kim K, Kwon I C, Ryu J H. Epidermal growth factor (EGF)-based activatable probe for predicting therapeutic outcome of an EGF-based doxorubicin prodrug. *J. Control Release.* 2020 Dec 10;328:222-236.

Pudełek M, Król K, Catapano J, Wróbel T, Czyż J, Ryszawy D. Epidermal Growth Factor (EGF) Augments the Invasive Potential of Human Glioblastoma Multiforme Cells via the Activation of Collaborative EGFR/ROS-Dependent Signaling. *Int. J. Mol. Sci.* 2020 May 20;21(10):3605.

Richani D, Gilchrist R B. The epidermal growth factor network: role in oocyte growth, maturation and developmental competence. *Hum. Reprod. Update.* 2018 Jan 1;24(1):1-14.

# Chapter 8

# Estrogen and Progesterone Receptors

**Karolina Siedlec**
**Dorota Bartusik-Aebisher**
**and David Aebisher***
Medical College of the University of Rzeszów, Poland

## Abstract

Estrogen and progesterone are sex hormones synthesized from cholesterol and therefore they belong to the steroid hormones. Estrogen and progesterone regulate the body's work by acting on various types of receptors. They are most often associated with the female reproductive cycle because they are essential for its proper course. Research on individual types of receptors is very promising, as these receptors could be used to prevent and inhibit many pathologies, such as endometriosis or cancers of the reproductive system.

**Keywords:** estrogen, progesterone receptor, hormone response elements (HREs), sex hormone binding glycoprotein (SHBG), tyrosine, lipid kinases

Estrogen and progesterone are sex hormones synthesized from cholesterol and therefore belong to the steroid hormones. They are most often associated with the female reproductive cycle because they are essential for its proper course. Initially, only the genomic way of interaction of these hormone receptors was

* Corresponding Author's Email: dbartusikaebisher@ur.edu.pl.

In: The Medical Biology Guide to Proteins
Editor: David Aebisher
ISBN: 979-8-88697-910-7

known, which resulted in proliferation, differentiation or secretion of specific biological substances by target cells, i.e., those containing steroid hormone receptors. In recent years, other complex modes of action of these receptors have been discovered. It has been found that the action of these receptors may vary in nature and, in addition to affecting gene transcription, also via a non-genomic route. By describing these mechanisms at the molecular level, it is possible to better understand them, and thus the potential use of this information for therapeutic purposes in the event of pathologies related to the functioning of given receptors. More and more studies are looking at ERβ receptors for their role in the treatment of breast cancer. Currently, steroid hormone receptors are divided into 3 groups: A - estradiol ERα and ERβ receptors, B - orphan estradiol receptors ERRα and ERRβ (they do not have their ligands), C - receptors for other steroid hormones, including progesterone PRA and PRB receptors. Each of the above-mentioned receptors consists of 5 domains with different functions (Khan et al. 2020).

The mentioned receptors have a similar structure, which consists of 5 modules, which are shown in Fig. 1. The N-terminal domain is the site of attachment of co-activation complexes, operating thanks to the present AF-1 sequence. Then there are two middle domains: the DNA-binding C-domain and the ligand-binding D-domain. In between there is a hinge region that is characterized by high sequence variability. It has not yet been fully explored. We further distinguish the E and C-terminal domains, in which the AF-2 sequence, responsible for the activation of transcription, is located. Estrogen receptors are encoded by separate genes, unlike progesterone receptors, which arise through alternative splicing of one gene (Medina et al. 2020).

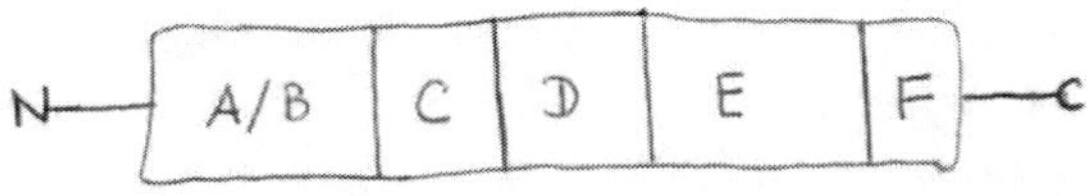


**Figure 1.** Domains building estrogen and progesterone receptors (own elaboration).

A/B - ligand independent transcription activation
C - attachment to DNA
D - hinge region
E - ligand attachment site
F - ligand dependent activation of transcription

Receptors dimerize, bind the ligand, and the active complex binds to special stretches of DNA called HREs (hormone response elements). HREs are located in close proximity to the target gene promoter. Another mode of action is the formation of heterodimers (αβ) with retinoid X receptors, and such a complex attach directly to HREs. In turn, the previously mentioned orphan receptors attach as homodimers (α or β) directly to the palindromic sequences of HREs, or as monomers to HREs on one side only. The function of receptors requires the presence of transcriptional factors and coregulators that activate pre-initiation complexes so that specific genes can be transcribed. At the same time, accompanying proteins strengthen the hormone-receptor complex and can affect the structure of chromatin, facilitating access to a given section of DNA and thus enabling gene transcription (Rozza-de-Menezes et al. 2021).

Hormones in the blood are transported in a specific way: in the case of estradiol, they are SHBG (sex hormone binding glycoprotein) and albumin, while progesterone is transported with the transcortin. Hormones penetrate target cells where they bind to receptors present in the cytoplasm or in the nucleus. It is worth noting that the location of receptors is variable, but most often it is determined by specific signals of nuclear location. The hormones attach to the receptors to take the place of the chaperones that protect the receptor when it is inactive. The ready hormone-receptor complex is translocated to the nucleus where it acts on the HREs. The transcription of genes by the action of polymerase II leads to a long-lasting cell response (Rozza-de-Menezes et al. 2021).

An interesting challenge for modern science seems to be the genomic mechanism of the action of hormones through receptors present in the cell membrane. This type of receptors has been distinguished for both estrogen and progesterone, but the multitude of steroid isoforms is surprising in the case of progesterone. Studies have shown that such receptors can translocate to the nucleus, which may indicate their effect on gene transcription. This type of receptors has been the least understood so far (Sporikova et al. 2018).

Another way steroid hormones work is fast, non-genomic. Its effect is much faster than in the case of genomic mechanisms. When the hormone binds to the membrane receptor, a cascade of cellular processes is immediately triggered, which in the end, it is worth noting, may also change the expression of genes in the cell (but not directly). This mechanism was related to, inter alia, with the regulation of ion channels or with the activation of G protein-related receptors. The effect on the cellular cascades of kinases: mitogen-activated (MAPK), tyrosine and lipid kinases was also described. The results of research conducted by Jakacka et al. (In mice) indicate that the proper functioning of the cell requires the work of both steroid nuclear receptors and membrane receptors. Research on membrane receptors gives hope for using them for anti-cancer activity (Tray et al. 2019).

In summary, estrogen and progesterone regulate the body's work by acting on various types of receptors. These receptors, despite a similar structure pattern, differ from each other in the mechanism of action. This results in a variable duration of the effects in the target cells as well as a variable duration. Research on individual types of receptors is very promising, as these receptors could be used to prevent and inhibit many pathologies, such as endometriosis or cancers of the reproductive system. The possible connections between the work of different types of receptors are also checked, so that the malfunctioning of one can be replaced by the increased work of the other. The variable location of the membrane receptors, which, thanks to nuclear localization signaling proteins, can move from the membrane to the nucleus is also unclear. There are suspicions that they affect gene expression. The percentage distribution of such receptors also raises questions as to why many of them locate at the nucleus, and what factors cause them to move towards the membrane. Expanding and then using knowledge about receptors can bring many benefits in modern methods of treatment, as evidenced by the number of discoveries made over the last 10 years.

## References

Khan ME, Haider G, Ahmed K, Ahmed M, Manzoor A, Masood A, Shaikh S, Abbas K. Ethnic differences in the receptors status of estrogen, progesterone and Her2/Neu among breast cancer women: A single institution experience. *J Pak Med Assoc.* 2020 Nov;70(11):1970-1974.

Medina MA, Oza G, Sharma A, Arriaga LG, Hernández Hernández JM, Rotello VM, Ramirez JT. Triple-Negative Breast Cancer: A Review of Conventional and Advanced Therapeutic Strategies. *Int J Environ Res Public Health.* 2020 Mar 20;17(6):2078.

Rozza-de-Menezes RE, Almeida LM, Andrade-Losso RM, de Souza Vieira G, Siqueira OHK, Brum CI, Riccardi VM, Cunha KS. A Clinicopathologic Study on the Role of Estrogen, Progesterone, and Their Classical and Nonclassical Receptors in Cutaneous Neurofibromas of Individuals With Neurofibromatosis 1. *Am J Clin Pathol.* 2021 Apr 26;155(5):738-747.

Sporikova Z, Koudelakova V, Trojanec R, Hajduch M. Genetic Markers in Triple-Negative Breast Cancer. *Clin Breast Cancer.* 2018 Oct;18(5):e841-e850.

Tray N, Taff J, Adams S. Therapeutic landscape of metaplastic breast cancer. *Cancer Treat Rev.* 2019 Sep;79:101888.

# Chapter 9

# Xanthine Oxidase

**Kornelia Rusek**
**Dorota Bartusik-Aebisher***
**and David Aebisher**
Medical College of the University of Rzeszów, Poland

## Abstract

Xanthine oxidase is the major protein involved in the formation of uric acid during purine metabolism. Its deficiency or excess is noticeable thanks to the determination of the concentration of purine. Xanthine oxidase is often found in the lungs, liver and serum. It plays a very important role in the metabolism of purines, and additionally catalyzes the reactions of caffeine metabolism, drugs, and the biosynthesis of second-order metabolites. It also significantly influences the process of oxidative stress. This study presents the negative effects of this protein by reducing the amount of reactive oxygen species. This research presents molecules with properties to reduce the number of ROS by binding them or eliminating their negative effects, but not acting directly on the reaction of uric acid formation.

**Keywords**: xanthine oxidase (XO), flavin adenine dinucleotide (FAD), flavonoprotein, oxidative stress

Xanthine oxidase (XO) was first described in 1902 thanks to the isolation of Bos taurus (domestic cattle) from milk. Most often it is found in the lungs,

* Corresponding Author's Email: dbartusikaebisher@ur.edu.pl.

In: The Medical Biology Guide to Proteins
Editor: David Aebisher
ISBN: 979-8-88697-910-7

liver and serum. It plays a very important role in the metabolism of purines, and additionally catalyzes the reactions of caffeine metabolism, drugs, and the biosynthesis of second-order metabolites. It also significantly influences the process of oxidative stress. Xanthine oxidase is synthesized from xanthine dehydrogenase, and this process can occur by oxidation of thiol residues or by peptide bond hydrolysis in the presence of proteases. It is an enzyme with the number EC 1.17.3.2, which classifies it as an oxyreductase, acting on the CH or CH2 group, with oxygen as an acceptor. This protein is relatively large, its molecular weight is 148.95 kDa. The co-factor of xanthyl oxidase is molybdenopterin, which has a molybdenum atom in its structure. It consists of 3 subunits: A, B, C, with a sequence length of 219, 350 and 763, respectively. Ligands bind to each constituent polypeptide chain. FES, i.e., the Fe2S2 compound, combines with the A subunit, FAD (flavin adenine dinucleotide) with the B subunit, and MTE, MOS, SAL and glycerol with the C subunit. Due to the structure of xanthine oxidase, it can be described as iron-molybdenum flavonoprotein containing [2Fe -2S]. Its enzymatic activity is determined by the –SH groups. This form of the enzyme was determined with the partially modified Nishino method (Anjum et al., 2018). The most important function of xanthine oxidase is related to purine metabolism. The reaction takes place in two stages. In the first step, hypoxanthine is oxidized to xanthine and hydrogen peroxide in the presence of water and oxygen. In this process, the hydrogen on the second carbon is replaced with oxygen. This results in the formation of 2,6-dihydroxypurine from 6-hydroxypurine. The second step is the oxidation of xanthine in the presence of water and oxygen to uric acid and hydrogen peroxide. Here, the hydrogen of the eighth carbon is replaced by oxygen: 2,6,8-trioxypurine is formed from 2,6-dihydroxypurine (De Giuseppe et al., 2029).

Another effect of xanthine oxidase is the reduction of cytochrome c. This happens indirectly, through a by-product of the enzyme's action: superoxide anion, which is one of the types of reactive oxygen species. This compound has the ability to react with nitric oxide, resulting in the formation of peroxynitrite, which has strong oxidizing properties.

$$O2- + NO \longrightarrow OONO-$$

The method for measuring the activity of XO is based on measuring the amount of uric acid produced from a given amount of xanthine or hypoxanthine. The measurement is performed in a pH = 7.4 buffer (usually carbonate or phosphate), at a temperature of 25 °C or 37 °C, with an incubation period of 15 to 30 minutes (depending on the author of the study). This test can be very useful in diagnosis, as high XO activity produces large amounts of uric acid. Presumably, the high concentration of this compound may be related to vascular problems in patients with hypertension, diabetes or cardiovascular diseases. In this case, XO inhibitors should be introduced (Guest et al., 2021).

In cases of too high uric acid concentration, xanthine oxidase inhibitors are used. An example of such a compound is alopurinol. It is an analogue of hypoxanthine, thanks to which it is oxidized to alloxanthin while reducing molybdenum. The entire reaction is catalyzed by xanthine oxidase, which is blocked by alloxanthin, preventing the enzyme from working further. Unfortunately, accumulating alopurinol can be harmful to health, so febuxostat and topiroxostat are more often recommended. These compounds additionally bind the inactive form of the enzyme, which prevents accumulation and reduces their activity. Additionally, it has been shown that oxipurinol has an inhibitory effect on xanthine oxidase. Its action prevents the production of large amounts of reactive oxygen species. In studies in rats with arterial hypertension, it was shown to lower the blood pressure. The above inhibitors, by blocking the action of XO, result in an increase in xanthine levels, which may result in xanthinuria, so it is important to check the condition of the patient's kidneys before taking them (Guest et al., 2021).

Disturbance of pro-oxidative and antioxidant balance resulting in excessive production of ROS may cause the so-called oxidative stress (or oxygen stress). In this process, the components of the cell, e.g., proteins, nucleic acids, lipids, can be damaged by oxidation. Interestingly, the increased production of ROS takes place in the joints: when we move, we deal with hypoxia of the joint tissues, and in the resting phase, reperfusion (i.e., restoration of blood flow after hypoxia), during which a large amount of reactive oxygen species is produced. This phenomenon is caused by xanthine oxidase, which catalyzes the hypoxanthine present in tissues and body fluids. During the oxidation reaction and the resulting reactions, reactive oxygen species are formed: hydrogen peroxide $H_2O_2$, superoxide anion $O_2$- and peroxynitrite OONO. They can cause autooxidation reactions of compounds such as hemoglobin, thiol compounds or catecholamines. To prevent their harmful effects, the antioxidant system is activated, which includes, among others, catalase, superoxide dismutase, glutathione peroxidase and glutathione reductase, which break down ROS (Nehlig et al., 2018).

In summary, xanthine oxidase is the major protein involved in the formation of uric acid during purine metabolism. Its deficiency or excess is noticeable precisely thanks to the determination of the concentration of this purine. During this reaction, large amounts of reactive oxygen species such as hydrogen peroxide and superoxide anion are released. In order to reduce the amount of ROS, inhibitors are administered which bind to the active site of xanthine oxidase. The best known are alopurinol (cumulative toxic), febuxostat and topiroxostat. High concentrations of this compound are likely to occur in cardiovascular disease, diabetes and hypertension, but this is an issue that is under constant research.

Xanthine oxidase is also being investigated for its role in oxidative stress. Such studies can overcome the negative effects of this protein by reducing the amount of reactive oxygen species produced. Due to the prevalence of oxygen stress in the society and interesting results concerning the presence of high concentrations of this enzyme in the above-mentioned diseases, the topic is under development and soon we can expect new inhibitors acting on xanthine oxidase. The aim of such research will probably be to find compounds reducing the number of ROS by binding them or eliminating their negative effects, but not acting directly on the reaction of uric acid formation.

## References

Anjum I, Jaffery S S, Fayyaz M, Wajid A, Ans A H. Sugar Beverages and Dietary Sodas Impact on Brain Health: A Mini Literature Review. *Cureus.* 2018 Jun 7;10(6):e2756.

De Giuseppe R, Di Napoli I, Granata F, Mottolese A, Cena H. Caffeine and blood pressure: a critical review perspective. *Nutr. Res. Rev.* 2019 Dec;32(2):169-175.

Guest N, Corey P, Vescovi J, El-Sohemy A. Caffeine, CYP1A2 Genotype, and Endurance Performance in Athletes. *Med. Sci. Sports Exerc.* 2018 Aug;50(8):1570-1578.

Guest NS, VanDusseldorp TA, Nelson MT, Grgic J, Schoenfeld BJ, Jenkins NDM, Arent SM, Antonio J, Stout J R, Trexler E T, Smith-Ryan AE, Goldstein ER, Kalman DS, Campbell BI. International society of sports nutrition position stand: caffeine and exercise performance. *J. Int. Soc. Sports Nutr.* 2021 Jan 2;18(1):1.

Nehlig A. Interindividual Differences in Caffeine Metabolism and Factors Driving Caffeine Consumption. *Pharmacol. Rev.* 2018 Apr;70(2):384-411.

# Chapter 10

# Selectin

**Klaudia Kuś**
**Dorota Bartusik-Aebisher***
**and David Aebisher**
Medical College of the University of Rzeszów, Rzeszów, Poland

## Abstract

The presented research shows the potential of selectins in the diagnosis of many diseases. Scientists conducting research and then publishing them in articles focused in particular on diabetes, cancer, atherosclerosis, and myocardial infarction. In their hypotheses, they assumed that there was a relationship between the concentration of selectins in the organisms of patients and their diseases. In recent years, researchers have published few articles on selectins.

**Keywords:** selectins, breast cancer, diabetes, lymphoblastic, chemotherapy

The aim of this study is proteins - selectins. The term covers a group of proteins involved in the inflammatory response. There are three main types of selectins: L-selectin, P-selectin and E-selectin. They are characterized by a similar building plan. They consist of three domains (lectin, EGF-like, CR) and a transmembrane and intracellular fragment. It is worth mentioning that the number of CR domains depends on the type of selectin. These proteins are

* Corresponding Author's Email: dbartusikaebisher@ur.edu.pl.

In: The Medical Biology Guide to Proteins
Editor: David Aebisher
ISBN: 979-8-88697-910-7

found on the surface of white blood cells and endothelial cells. Selectin binds to various proteins, most of which are mucins (Abadier et al., 2017).

L-selectin is the smallest of the selectins as it contains only two CR domains. This protein is found on the surface of leukocytes. Its function is to select the appropriate lymphocytes to remain in the lymph nodes. It also takes part in the transport of neutrophils to the place where the inflammatory reaction takes place. During inflammation, they cooperate with P selectin. P-selectin is the largest of the selectins. It occurs in the granules of blood platelets and in endothelial cells. It plays a major role in the initial stages of the inflammatory process. It also influences the process of blood clotting and neoplasms. E-selectin has six CR domains (Ivetic et al., 2018).

L-selectin is an indicator of the prevalence of diabetes mellitus. It has been shown that the expression of L-selectin is higher in patients than in healthy people and does not depend on the duration of the disease. These studies included a group of people with diabetes for less than five years, more than five years, and a group of healthy people. The expression of the adhesion molecules was analyzed by three-color flow cytometry.

On the other hand, studies on the population size of Treg lymphocytes expressing L-selectins (CD62L) in diabetic patients show that it is lower than in healthy subjects. Thus, a lower percentage of Tregs (CD62L) may be the cause of the development of inflammation in patients with type I diabetes, because (CD62L) plays an important role in the colonization of tregs, e.g., in lymph nodes. On thc other hand, studies on NOD mice show that lower expression of L-selectin-expressing T cells causes diabetes and only Treg (CD62L) can protect against this disease (Ivetic et al., 2019).

Studies were also carried out in cancer patients and correlations with the concentration of L-selectin were also carried out. In breast cancer patients, L-selectin levels have been tested as a measure of the activity of granulocytes that kill cancer cells. The concentration of L-selectin in patients was reduced both before and during chemotherapy. These results were derived from sensitive ELISA sandwich systems. Similar conclusions were shown when examining patients with endometrial cancer. The concentration of L-selectin was decreased, so these results indicate a reduced ability of granulocytes to bind to tumor cells. L-selectin expression was also measured in lung cancer patients. It has been shown that in these people an increased concentration of L-selectins is observed. These results were explained by the fact that the higher concentration of these molecules may be influenced by the activation of neutrophils by pro-inflammatory cytokines (Rahman et al., 2021).

The effect of L-selectin levels in patients with acute myeloblastic leukemia and acute lymphoblastic leukemia was investigated. The sL level was determined by ELISA. The presented results indicated that the concentration of L-selectin in all patients with acute leukemia was higher than in the control group. It was found that the analyzed L-selectin level can be used in the diagnosis of ALL and AML leukemia activity. However, in other studies it was concluded that the increase in L selectin concentration increases in leukemias, especially in the M2 subtype of acute myeloid leukemia (Sakuma et al., 2020).

P - selectin activates the processes of adhesion of leukocytes to the endothelium in inflammation. The result of this process is diapedesis from the vessel to the inflammatory sites. Disturbances in the amount of P-selectin impairs the rolling process of leukocytes along the endothelium. The increase in P-selectin expression was observed in atherosclerotic lesions. On the other hand, the lack of P-selectin can reduce the process of atherosclerotic plaque formation. In the conducted studies, it was found that P-selectin plays an important role in processes related to thrombosis of blood vessels, but also in coronary disease or myocardial infarction. The role of P-selectins is to participate in the processes of connecting platelets with leukocytes. P-selectin leads to the formation of aggregates which result in the formation of a clot. The results of blood laboratory tests in patients with myocardial infarction show that in these patients an increased concentration of P-selectin is observed. These studies promise to be a reliable measurement for the rapid diagnosis of myocardial infarction. E - selectins are also involved in causing inflammation. In patients with enteritis, an increase in their concentration is observed (Sakuma et al., 2020).

The presented research shows the potential of selectins in the diagnosis of many diseases. Scientists conducting research and then publishing them in articles focused in particular on diabetes, cancer, atherosclerosis, and myocardial infarction. In their hypotheses, they assumed that there was a relationship between the concentration of selectins in the organisms of patients and their diseases. The results and the conclusions reached confirmed the assumptions. It was also noticed that depending on the disease entity, other types of P, L, E selectin changes its concentration. L-selectin has been used in the diagnosis of diabetes and neoplasms, and P-selectin in the formation of atherosclerotic plaques. In addition to diagnosis, the observation of selectins may be helpful in understanding the mechanisms of the immune system during the disease process. Most of these studies were performed before 2010. In recent years, researchers have published few articles on selectins. The identification of L-selectins in particular in the course of neoplastic diseases raises hopes for their wider application in everyday practice.

## References

Abadier M, Ley K. P-selectin glycoprotein ligand-1 in T cells. *Curr. Opin. Hematol.* 2017 May;24(3):265-273.

Ivetic A, Hoskins Green H L, Hart S J. L-selectin: A Major Regulator of Leukocyte Adhesion, Migration and Signaling. *Front. Immunol.* 2019 May 14;10:1068.

Ivetic A. A head-to-tail view of L-selectin and its impact on neutrophil behaviour. *Cell Tissue Res.* 2018 Mar;371(3):437-453.

Rahman I, Collado Sánchez A, Davies J, Rzeniewicz K, Abukscem S, Joachim J, Hoskins Green HL, Killock D, Sanz MJ, Charras G, Parsons M, Ivetic A. L-selectin regulates human neutrophil transendothelial migration. *J. Cell Sci.* 2021 Feb 8;134(3): jcs250340.

Sakuma K, Kannagi R. Selectin-Binding Assay by Flow Cytometry. *Methods Mol. Biol.* 2020; 2132:111-118.

## Chapter 11

# Integrin

**Wojciech Szynal**
**Dorota Bartusik-Aebisher***
**and David Aebisher**
Medical College of the University of Rzeszów, Rzeszów, Poland

### Abstract

Integrins are a group of heterodimers in vertebrates consisting of 24 different transmembrane receptors. Integrins are the most important group among cell adhesion receptors, and also one of the most numerous groups of external membrane receptors. They are present virtually in every cell of the body, except for red blood cells. The main focus of this chapter was on their structural and biochemical features. The studies were mainly conducted using antibodies and the ligand binding mechanism in terms of mutations. In the cellular signals still hide many unknowns.

**Keywords:** integrins, leukocytes, integrin-associated kinase (IKL), cytoskeleton, intracellularly, extracellularly

Integrins are a group of heterodimers in vertebrates consisting of 24 different transmembrane receptors. They are composed of α and β subunits. There are 18 α and 8 β subunits. Integrins are the most important group among cell adhesion receptors, and also one of the most numerous groups of external

* Corresponding Author's Email: dbartusikaebisher@ur.edu.pl.

In: The Medical Biology Guide to Proteins
Editor: David Aebisher
ISBN: 979-8-88697-910-7

membrane receptors. They are present in virtually every cell of the body, except for red blood cells. Their name comes from the function they were originally thought to perform, that is, connecting the extracellular substance with the actin of the cell cytoskeleton (Boppart et al., 2019).

They play an important role in many important processes, such as: conduction of intracellular and extracellular signals, cell proliferation and their organization into tissues. For this reason, any errors in the regulation of integrin activity may lead to the development of many dangerous diseases, including cancer, in the fight against which today a solution is being sought precisely in changing the activity of these receptors. This regulation can take place at various stages, including ligand attachment and intracellular protein binding. The discovery of the importance of integrins for the functioning of multicellular organisms led to intensified work on them, which led to the understanding of the key role of many signaling proteins, adapter proteins and cellular mechanisms in this process. Ultimately, it turns out that the regulation of integrin functions is very complex and closely related to signaling mechanisms and the cytoskeleton of the cell (Cooper et al., 2019).

**Table 1.** Classification of integrins

| | Integrins β1 | Integrins β2 | Integrins αV |
|---|---|---|---|
| Specific receptors for leukocytes | β1α9 | β2αL | |
| | β1α4β7αE | β2αM | |
| | | β2αX | |
| | | β2αD | |
| RGD sequence receptors | β1α5 | | αVβ8 |
| | β1α8 | | αVβ6 |
| | | | αVβ5 |
| | | | αVβ3αIIb |
| Collagen receptors | β1α1 | | |
| | β1α2 | | |
| | β1α10 | | |
| | β1α11 | | |
| Laminin receptors | β1α3 | | |
| | β1α7 | | |
| | β1α6β4 | | |

Source: Own elaboration.

As mentioned before, integrins are heterodimers composed of non-covalently linked α and β subunits, among which there are 18 α and 8 β subunits that can form 24 different heterodimers. Integrins can be grouped into subgroups because of their ligand binding properties derived from the arrangement of the subunits. In such a classification, three largest representatives can be distinguished: β1 integrins, β2 integrins and αV integrins. Likewise, the β subunits contain partially the same fragments within their group (Moreno-Layseca et al., 2019).

The extracellular part of an integrin consists of a head (mainly the α subunit) connecting ligands (components of the intercellular substance, including collagen, laminin, fibronectin) connected to two legs (β and α subunit), each of which connects to the intracellular part. Each α and β subunit of integrins is composed of several domains linked by flexible connections. The subunit has a helix connecting it to the cell membrane and a short unstructured cytoplasmic tail (usually). Subunit sizes vary, but typically each β subunit contains about 1000 amino acids and the α subunit about 750 (Koivisto et al., 2018).

Integrins perform a significant number of functions for the cell, tissue, and thus the entire organism, but the two most important of them are: connecting the extracellular space with the cytoskeleton of the cell and the conduction of signals between the extracellular space and the cell, both intracellularly and extracellularly. The first discovered function of integrins (which is where their name comes from) was to link the extracellular space with the cytoskeleton. In most integrins, this connection is made with the actin part of the cytoskeleton, but even as in the case of α6β4, they can combine with intermediate filaments. Some of the components of this mechanical linkage, such as the protein talin, have a dual function in addition to being physically linked. Integrins act as nano-scale mechanoreceptors that transfer the tension between the cytoskeleton and the extracellular space in order to maintain the structural integrity of the tissue. The signals generated in this way affect cell physiology through a set of intracellular signaling mechanisms, including autocrine and paracrine ones. When we talk about integrins, we also mean signal receptors involved in extracellular and intracellular conduction. This is mainly of importance in activating integrins. Proteins such as talin, filamine, kindlin, migfilin, tyrosine kinase 2 (FAK), and integrin-associated kinase (IKL) may be involved in the regulation of integrin activation. After ligand binding, the integrin changes its conformation, resulting in further signal transduction. A complex signaling pathway is specific to each cell (Sun et al., 2019).

Other important functions of integrins are:

- Facilitating cell migration
- Impact on key cellular processes

- Contribute to many serious diseases when their structure and function are damaged (e.g., autoimmune diseases, cardiovascular diseases, fetal malformations)
- Facilitating the emergence and development of neoplasms in the event of structural abnormalities
- Cell-to-cell adhesion
- Extravasation
- As receptors for some viruses (adenoviruses, hantaviruses, foot-and-mouth disease virus, ECHO viruses, poliovirus and others)
- Participation in the formation of clots

Despite the fact that integrins were discovered more than 30 years ago, only recently has been an in-depth interest in and detailed description of their functions. The main focus was on their structural and biochemical features. The studies were mainly conducted using antibodies and the ligand binding mechanism in terms of mutations. Now we have the opportunity to look at integrins at the atomic level. Experimental and computer studies focusing on the interaction between integrins and the extracellular and intracellular space are very important, nevertheless, it would be worth using these agents in well-known and relevant in vitro and in vivo models, focusing on the mechanisms of development of various diseases and their relationship with integrins. We still do not know how integrins enable a cell to interpret the binding of different ligands differently and exactly how the deactivation process looks like. These are just a few of the many issues that still need to be clarified. Let us hope that the coming years will allow us to understand a little more about the number of processes in which integrins are involved.

## References

Boppart M D, Mahmassani Z S. Integrin signaling: linking mechanical stimulation to skeletal muscle hypertrophy. *Am. J. Physiol. Cell Physiol.* 2019 Oct 1;317(4):C629-C641.

Cooper J, Giancotti F G. Integrin Signaling in Cancer: Mechanotransduction, Stemness, Epithelial Plasticity, and Therapeutic Resistance. *Cancer Cell.* 2019 Mar 18;35(3):347-367.

Koivisto L, Bi J, Häkkinen L, Larjava H. Integrin αvβ6: Structure, function and role in health and disease. *Int. J. Biochem. Cell Biol.* 2018 Jun;99:186-196.

Moreno-Layseca P, Icha J, Hamidi H, Ivaska J. Integrin trafficking in cells and tissues. *Nat. Cell Biol.* 2019 Feb;21(2):122-132.

Sun Z, Costell M, Fässler R. Integrin activation by talin, kindlin and mechanical forces. *Nat. Cell Biol.* 2019 Jan;21(1):25-31.

# Chapter 12

# Coronin

**Natalia Wójcik**
**Dorota Bartusik-Aebisher***
**and David Aebisher**
Medical College of the University of Rzeszów, Poland

## Abstract

Coronins are conserved, low-evolving proteins that play an important role in the immune system. Members of the coronin protein family are important regulators of the actin cytoskeleton and are related to actin. Coronins functions have not yet been fully described. It is known, however, that these proteins are important in the functioning of immunity, especially natural immunity.

**Keywords:** coronins, autoinflammatory diseases, genetic disease, neutrophil, macrophage

Coronins are proteins that play an important role in the immune system. Coronins affect the activity and functioning of cells, and thus participate indirectly in the processes of natural and acquired immunity. The mechanism by which coronins connect with the cell membrane is based on their interaction with cholesterol, since they do not have transmembrane domains. So far, 7 proteins have been classified into the coronin family (Chen et al., 2020). Members of the coronin protein family are important regulators of the actin

---

* Corresponding Author's Email: dbartusikaebisher@ur.edu.pl.

In: The Medical Biology Guide to Proteins
Editor: David Aebisher
ISBN: 979-8-88697-910-7

cytoskeleton and are related to actin. Their presence has been demonstrated as a component that binds to actin-myosin complexes. This fact was described by de Hostosa and his colleagues in 1991 by isolating these proteins from the amoeba Dictyostelium discoideum. The presence of coronin 7 has been proven in the Golgi apparatus. This leads to the conclusion that it has an exceptional effect not on the cytoskeleton but on the structure of internal membranes. It is also important that mutations in CORO1A, encoding coronin-1A, the dominant coronin in lymphocytes, cause various degrees of T-cell lymphopenia, susceptibility to infections and dysregulation of immunity. It follows that coroninas play an important role in bacterial and viral infections and in diseases of the genetic background. Coronin plants are therefore an element of maintaining homeostasis in the body (Chen et al., 2020).

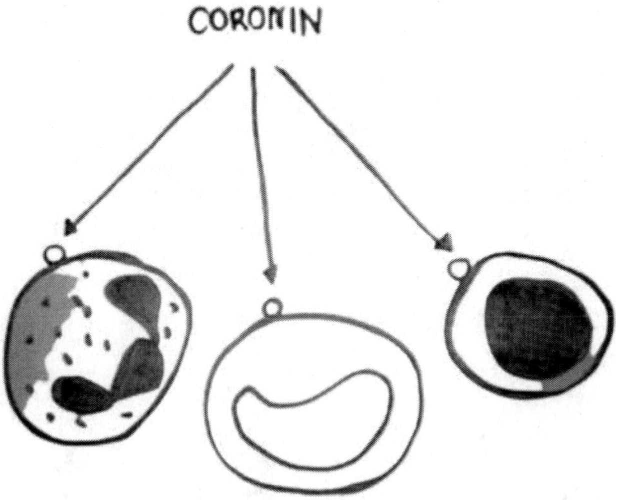


**Figure 1.** Cells with coronin membranes: neutrophil, macrophage and lymphocyte (own elaboration).

As already mentioned, coronation 1 was discovered by de Hostos in 1991. Scientists isolated it from the amoeba Dictyostelium discoideum. Several similarly structured coronets were later discovered in mammals. At that time, the focus was on the role of these proteins in rebuilding the skeleton. Later, further functions were discovered, including the role in calcium signaling. Mammalian coronin studies were conducted in mice. It was then that the influence of coronin on immune cells, T and B lymphocytes was discovered (Ford et al., 2019).

Coronins contain many repeating sequences of 40 amino acids that have "WD repeats" (tryptophan W and asparagine D) similar to the beta subunits of G proteins. Koronin 1A, which is the best known of these proteins, is made up of a short chain with an N-terminal segment. This segment creates 7 WD repeats and mediates the binding of the plasma membrane. The centrally located linker is the F-actin binding site. The C-terminal fragment mediates homotrimerization and association with the cytoskeleton. This structure implies that coronin 1 can bind the plasma membrane to the cytoskeleton and induce its remodeling in response to extracellular signals (Kim et al., 2020).

The role of coronin 1 is assigned to the generation of naïve effector and memory T cells, T cells, and B cells. Studies have shown that the absence of coronin 1 is associated with increased apoptosis-related lymphocyte mortality. The mechanism of this function is as follows. The lack of coronin and its presence in a mutated form conditions incorrect mobilization of calcium ions. As a result, the cascade activation process of the TCR receptor and apoptosis of T cells are inhibited. The lack of protein also leads to the preservation of effector T cells and memory for longer, which proves the important role this protein plays in the mechanism of binding to the TCR receptor. Koronina 1 is also involved in bacterial infections. Its role has been demonstrated when the macrorganism is infected with the microorganism Mycobacterium tuberculosis. This protein blocks the delivery of the phagosomes with the microorganism to the lysosomes in the macrophages and thus the mycobacteria are not destroyed. The same situation was observed for M.leprae and Helicobacter pylori. Based on these facts, it was concluded that coronin 1 plays a key role in macrophage phagocytosis and hence non-specific immunity. Coronin is also associated with macrophages in the activation of calcium signaling (Martorella et al., 2017).

The uptake of microorganisms is associated with a transient increase in calcium concentration in the cell, and its signaling is conditioned by coronin 1. As far as mast cells are concerned, the influence of coronin on their function is ambiguous. Koronina 1 also plays a role in viral infections. They have so far been described in mice infected with lymphocytic meningitis virus and vesicular stomatitis virus. Coronin-deficient 1 mice are more susceptible to infection with the vesicular stomatitis virus. They also have less B-cell activity in the production of antibodies, which leads to fewer antibodies. Coronin 1 in mice also plays a role in genetic disease. In mice, the deficiency of this protein led to encephalomyelitis. This is associated with the abnormal activation of SMAD3 and MAD in T17 lymphocytes. In addition, a lack of coronin 1 on neutrophils has been observed in patients with cystic fibrosis, which also binds the protein to this disease (Tagliatela et al., 2020).

Summarizing the function of coronin proteins, it can be stated with certainty that they play many important roles. They should definitely be related to the cytoskeleton - actin filaments, because this fact was described as the first in terms of discoveries concerning coronin. They are relatively recently discovered, and by 2021, 306 studies on coroninitis have been published. So they are not among the most popular among scientists. Since the discovery of coronets in 1991, significant research has been carried out to elucidate their molecular structure and cellular mechanisms of action. Nevertheless, their functions have not yet been fully described. It is known, however, that these proteins are important in the functioning of immunity, especially natural immunity. Other works concern their participation in cancer and genetics. Therefore, it is worth researching and learning about the exact functioning of these proteins so that they can be used in potential disease therapy. With new evidence on the potential of anti-Coronin-1A monoclonal antibodies in the treatment of B-cell cancers and T-cell-mediated autoinflammatory diseases, it can be concluded that coronin-1A is involved in the overall regulation of the immune system and its abnormal expression may lead to immunodeficiency or autoimmunity.

## References

Chen Y, Xu J, Zhang Y, Ma S, Yi W, Liu S, Yu X, Wang J, Chen Y. Coronin 2B regulates dendrite outgrowth by modulating actin dynamics. *FEBS Lett.* 2020 Sep;594(18): 2975-2987.

Ford M L. Coronin-1, King of Alloimmunity. *Immunity.* 2019 Jan 15;50(1):3-5.

Kim G Y, Lim H J, Kim W H, Park H Y. Coronin 1B regulates the TNFα-induced apoptosis of HUVECs by mediating the interaction between TRADD and FADD. *Biochem. Biophys. Res. Commun.* 2020 Jun 11;526(4):999-1004.

Martorella M, Barford K, Winkler B, Deppmann CD. Emergent Role of Coronin-1a in Neuronal Signaling. *Vitam. Horm.* 2017;104:113-131.

Tagliatela A C, Hempstead S C, Hibshman P S, Hockenberry M A, Brighton HE, Pecot C V, Bear J E. Coronin 1C inhibits melanoma metastasis through regulation of MT1-MMP-containing extracellular vesicle secretion. *Sci. Rep.* 2020 Jul 20;10(1):11958.

# Chapter 13

# Ependymin

**Iga Serafin**
**Dorota Bartusik-Aebisher***
**and David Aebisher**
Medical College of the University of Rzeszów, Poland

## Abstract

Ependimin is a glycoprotein that was first discovered in osseous fish in the lining of the cavities of the central nervous system. The significant amount of ependymin is also found in the cerebrospinal and extracellular fluid in the brain of these animals. Ependimin is related to EPDR glycoproteins with a variety of functions. The structures of EPDR are similar in all organisms in which it occurs. The monomeric subunit consists of two antiparallel planes and a hydrophobic pocket, probably used for lipid attachment. Studies have been carried out to investigate the influence of the EPDR1 gene on the development of neoplasms: HCC, BLCA and CRC. It has been shown that in neoplastic cells the expression of this gene is higher than in healthy cells. The role of EPDR1 in the regulation of cancer resistance is related to the movement of immune cells to the vicinity of the tumor. It seems reasonable to conduct research on the use of the EPDR1 gene in medicine as a biomarker for cancer.

**Keywords:** ependimin (EPN or EPD), hepatocellular carcinoma (HCC), ependymin, progression-free survival (PFS), overall survival (OS)

---

* Corresponding Author's Email: dbartusikaebisher@ur.edu.pl.

In: The Medical Biology Guide to Proteins
Editor: David Aebisher
ISBN: 979-8-88697-910-7

Ependimin (EPN or EPD) is a glycoprotein that was first discovered in osseous fish in the lining of the cavities of the central nervous system. Its significant amount is also found in the cerebrospinal and extracellular fluid in the brain of these animals. Ependimin is a protein consisting of approximately 200 amino acids, among which 4 cysteines play an important structural role. This glycoprotein plays an important role in the nervous system. It has been shown that in bone skeletal fish it is involved in the mechanisms of long-term memory, neuron regeneration, calcium ion management in the brain, and also influences adaptation to cold and aggression. Later, the existence of ependiminin-related proteins (EPDR) has also been proven, occurring, inter alia, in in echinoderms, amphibians, and also in mammals, specifically mice (MERPS), and in humans, where they are abbreviated as UCC1-, they arise in cancerous cells of the colon. Currently, genes encoding ependiminin-related proteins are referred to as EPDR1. MERP1 proteins are also found in tissues outside the CNS, such as the brain, heart, skeletal muscle, kidney, and hemopoietic cells, as well as in several malignant cell lines (Chen et al., 2020). The conducted studies, the results of which are presented below, investigated the effect of EPDR1 on the cells of the immune system in the case of cancers, incl. hepatocellular carcinoma (HCC) and the potential carcinogenic role of this gene, including bladder cancer (BLCA). Moreover, the use of EPDR1 as a biomarker in oncological patients with selected neoplasms was considered (Deshmukh et al., 2019).

In the brain of goldfish, in which ependiminin was first detected, researchers have proven the existence of at least two transcripts that differ slightly in structure, but are responsible for the formation of similar precursor compounds for this protein. They condition the production of two kinds of ependymin- $\alpha$ and $\beta$. In the structure of the precursors, 216 amino acid residues were found with two possible N-linked glycosylation sites. The proteins associated with ependymin EPDR are found in many organisms, including mammals and amphibians. After examining the structures of these proteins taken from mice, frogs and humans, it was found that they are similar in each case - the monomeric subunit is made up of 2 antiparallel sheets with one-sided curvature, resulting in a hydrophobic pocket that is probably used to bind with fatty acids. Within the monomeric subunit there are 4-6 cysteine residues, forming 2-3 disulfide bridges inside the molecule (Deshmukh et al., 2019). There are at least 2 N-glycosylation sites in the structure of EPDR, but with a different location than in fish (there are 2 asparagine sites there). After crystallization, it was found that the 4 monomeric subunits of human EPDR connect with other subunits through 2 orthogonal double axes. Moreover, the fixation of the calcium ion in the water-amino acid network near the hydrophobic pocket was observed. The folded structure of EPDR1 resembles the fold of bacterial proteins: VioE as well as LolA and LolB, so it can be assumed that it can perform similar functions related to the presence of a hydrophobic pocket - attach the substrate taking the role of an enzyme and transport lipids (Gimeno-Valiente et al., 2020).

In recent years, many studies have focused on the correlation of EPDR1 gene expression with the occurrence and development of neoplasms, e.g., HCC hepatocellular carcinoma, BLCA bladder and CRC colon. In 2001, it was possible to detect a transcript in CRC cell lines that was not present in normal cells. Upon examination of its cDNA, similarity with the ependimin-UCC1 gene, now referred to as EPDR1, was noticed. In 2020, HCC took 6th place in terms of the number of cases and 2nd in terms of mortality among all malignant neoplasms in the world. Currently used diagnostic methods can be unreliable, which leads to delayed diagnosis, and thus to a worse prognosis for the patient. For this reason, it seems important to test EPDR1 in terms of its usefulness as a biomarker for the developing cancer. The study showed, using bioinformatic database analysis, that higher EPDR1 expression occurs in malignant tissues (HCC) than in healthy tissues, and that patients with more EPDR1 have significantly shorter overall survival (OS) as well as progression-free survival (PFS) and relapses (DSS) (McDougall et al., 2018). The relationship between the occurrence of EPDR1 and the activity of the immune system was investigated. The analysis showed that the increased expression of this gene correlated with the infiltration of immune cells into the tumor site, which indicates that the presence of EPDR1 may modulate cancer resistance. The researchers reached similar conclusions, pointing to a higher expression of EPDR1 in the cells of the colon changed with neoplasms as compared to the normal mucosa. The presence of this gene promotes tumor invasiveness, proliferation and cell migration. A similar effect was shown in the case of BLCA bladder cancer, where it was shown that patients with high EPDR1 expression statistically worse prognosis related to faster disease development and higher tumor malignancy and metastases (Park et al., 2019).

The presence of ependymin is mainly found in the central nervous system, where this protein participates in important processes such as regeneration of neurons, long-term memory mechanisms, and even behavior regulation. The structure of EPDR is similar in all organisms in which it occurs to the monomeric subunit and consists of two antiparallel planes. The structure is stabilized by disulfide bridges formed between 4 or 6 cysteine residues. Studies have been carried out to investigate the influence of the EPDR1 gene on the development of neoplasms: HCC, BLCA and CRC. It has been shown that in neoplastic cells the expression of this gene is higher than in healthy cells. In addition, the role of EPDR1 in the regulation of cancer resistance, related to the movement of immune cells to the vicinity of the tumor, was found. The correlation of more EPDR1 with poor prognosis of the course of the disease for a patient with HCC, related to shorter overall survival or relapse-free survival, was also indicated, as in the case of CRC and BLCA, where there was a correlation between more EPDR1 and more severe course of the disease, and faster its development, as a result of increased proliferation and migration of cancer cells, and thus a greater number of metastases.

## References

Chen R, Zhang Y. EPDR1 correlates with immune cell infiltration in hepatocellular carcinoma and can be used as a prognostic biomarker. *J. Cell Mol. Med.* 2020;24(20):12107-12118.

Deshmukh A S, Peijs L, Beaudry J L, Jespersen N Z, Nielsen C H, Ma T, Brunner A D, Larsen T J, Bayarri-Olmos R, Prabhakar B S, Helgstrand C, Severinsen M C K, Holst B, Kjaer A, Tang-Christensen M, Sanfridson A, Garred P, Privé G G, Pedersen B K, Gerhart-Hines Z, Nielsen S, Drucker D J, Mann M, Scheele C. Proteomics-Based Comparative Mapping of the Secretomes of Human Brown and White Adipocytes Reveals EPDR1 as a Novel Batokine. *Cell Metab.* 2019;30(5):963-975.e7.

Gimeno-Valiente F, Riffo-Campos Á L, Ayala G, Tarazona N, Gambardella V, Rodríguez FM, Huerta M, Martínez-Ciarpaglini C, Montón-Bueno J, Roselló S, Roda D, Cervantes A, Franco L, López-Rodas G, Castillo J. EPDR1 up-regulation in human colorectal cancer is related to staging and favours cell proliferation and invasiveness. *Sci. Rep.* 2020;10(1):3723.

McDougall C, Hammond M J, Dailey S C, Somorjai I M L, Cummins S F, Degnan B M. The evolution of ependymin-related proteins. *BMC Evol. Biol.* 2018;18(1):182.

Park J K, Kim K Y, Sim Y W, Kim Y I, Kim J K, Lee C, Han J, Kim C U, Lee J E, Park S. Structures of three ependymin-related proteins suggest their function as a hydrophobic molecule binder. *IUCrJ.* 2019;6(Pt 4):729-739.

# Chapter 14

# Cadherin

**Kamil Rachański**
**Dorota Bartusik-Aebisher***
**and David Aebisher**
Medical College of the University of Rzeszów, Rzeszów, Poland

## Abstract

Cadherins are transmembrane proteins involved in the formation of adhesive bonds between cells. Protein plays a key role in several important processes during embryogenesis, such as the formation of gastrula, neurula, and organogenesis. Adhesion based on this protein is necessary to maintain the proper architecture of tissues in organisms. Despite the fact that cadherins have been discovered for over 40 years, their development is still slow and requires significant financial outlays. However, their incorrect expression leads to oncogenesis and metastasis. Therefore, a better understanding of cadherins is critical to cancer clinical applications, especially as therapeutic targets. Understanding how N-cadherin influences cell behavior will enable the development of therapies to combat its activity and prevent cancer cell growth, invasion and metastasis. Adhesion molecules are promising new targets in the treatment of cancer but could also be useful in predicting patient prognosis in human and veterinary medicine.

**Keywords:** cadherins, polypeptides, desmocoline gene, fibroblast growth factor receptor (FGFR)

---

* Corresponding Author's Email: dbartusikaebisher@ur.edu.pl.

In: The Medical Biology Guide to Proteins
Editor: David Aebisher
ISBN: 979-8-88697-910-7

Cadherins are transmembrane proteins involved in the formation of adhesive bonds between cells. First described in the 1980s by Hyafil and Peyrieras as a result of studies on mice and chickens. The condition for the formation of interactions by cadherins are Ca2 + ions, therefore the presence of calcium-binding agents such as EDTA (edetic acid) in the environment, i.e., chelating compounds, causes the breakdown of these bonds. There are over 100 different types of cadherins that can be divided into four groups: classical, desmosomal, proto-adherin, and unconventional. Protein plays a key role in several important processes during embryogenesis, such as the formation of gastrula, neurula, and organogenesis (Cao et al. 2019). Adhesion based on this protein is necessary to maintain the proper architecture of tissues in organisms. In mature tissues, they are responsible for integrity, cell polarity maintenance, and homeostasis. Classic cadherins are inherently related to catenins with which they form cadherin-catenin complexes. During changes in the body's cell, cadherins are often inactivated or functionally blocked. An example is the epithelial E-cadherin belonging to the classical cadhedrins, the deletion of which in the murine mammary epithelium in the p53 gene promotes tumor initiation and progression to metastasis. Previous research has focused on the relationship between cadherins and malignant neoplasms. The resulting drugs targeting these proteins have been developed and tested in clinical trials. The aim will be to discuss the different types of cadherdrin and the effects of dysregulation of their expression in neoplastic diseases (Corso et al. 2020).

The cadherins can be divided into several subtypes: classical type I cadherins such as E-cadherin, N cadherin, P-cadherin, and classical type II cadherins such as VE cadherin, OB cadherin, desmosomal cadherins, protocadherins.

- E-cadherin (epithelial) is a key functional component of the adjacent junctions of all epithelial cells. It plays an important role in the creation and maintenance of intercellular adhesion, cell polarity, and tissue architecture. E-cadherins are found on the surface of epithelial cells and can bind to cadherins of the same type on others, forming bridges. The loss of E-cadherin cell adhesion molecules is causally related to the formation of epithelial neoplastic cells.
- Changes in any type of cadherin expression may not only control tumor cell adhesion, but may also affect signal transduction leading to uncontrolled growth of cancer cells.

- N-cadherin (nerve) determines the correct architecture of some tissues, but also plays a role in cell communication, ie it has a signaling function, is involved in the formation of synapses in neurons and in the formation of the vascular wall necessary for vascular stabilization. In N-cadherin, in contrast to E-cadherin, the migratory and invasive abilities of neoplastic cells are increased. In addition, during tumor progression, N-cadherin may affect cell survival, facilitate the EMT process, and facilitate migration / invasion by recruiting signaling molecules. Penetration of N-cadherin with another membrane protein, such as the fibroblast growth factor receptor (FGFR), also activates signaling cascades that affect cell proliferation, invasion, and cell-cell adhesion. In normal epithelial cells, N-cadherin is absent or low in expression.
- P-cadherin - This is the first time this adhesive molecule has been detected in the placenta of mice. Expression of β-cadherin has been shown to be restricted to the non-embryonic ectoderm and the visceral endoderm, the structures that make up the placenta. P-cadherin dysfunction is strongly associated with the formation of bladder, pancreas, breast, ovary, prostate, skin, stomach and colon cancers (Mendonsa et al. 2018).
- VE-cadherin (vaso-endothelial) is found mainly in endothelial cells, where it plays a key role in vascular integrity and permeability and enables homotypic adhesion between cells.
- OB-cadherin is responsible for homophilic cell adhesion and is involved in many other biological functions, including cytoskeleton organization, tissue morphogenesis, cell migration and invasion (Na et al. 2020).

*Protocadherin* (C-Pcdhs, NC-Pcdhs) The Pcdhs clustered proteins are detected throughout the neuronal soma, dendrites, axons, and observed at synapses and growth cones. Like classical cadherins, they mediate cell-cell adhesion and are also involved in the formation of homophilic trans interactions.

*Desmosomal cadherins* form intercellular junctions that tightly connect adjacent cells called desmosomes. The adhesion proteins belonging to the cadherins desmoglein and desmokolin are involved in this connection. Desmosomes are especially important for the integrity of the heart muscle and skin epithelium. For this reason, mutations in the desmocoline gene may affect the adhesion of cells subjected to mechanical stress, especially cardiomyocytes and keratinocytes (Mendonsa et al. 2018).

Cadherins are synthesized as polypeptides and undergo many post-translational modifications to become proteins that mediate cell adhesion and recognition. The polypeptides are approximately 720–750 amino acids in length. Each cadherin has a small C-terminal cytoplasmic component, a transmembrane component, and the remainder of the protein is outside the cell. The transmembrane component consists of single-chain repeats of the glycoprotein. In Figure 1, there is desmosomal cadherin that tightly connects adjacent cells with each other by means of snaps via intermediate filaments. The second compound, the classic type I cadherin, connects with the actin cytoskeleton on the surface through interactions with catenins in the cytoplasm, which allows it to be anchored in the cytoskeleton. In this way, a polypeptide present on the surface of one cell can bind to that on the other, forming bridges (Wong et al. 2018).

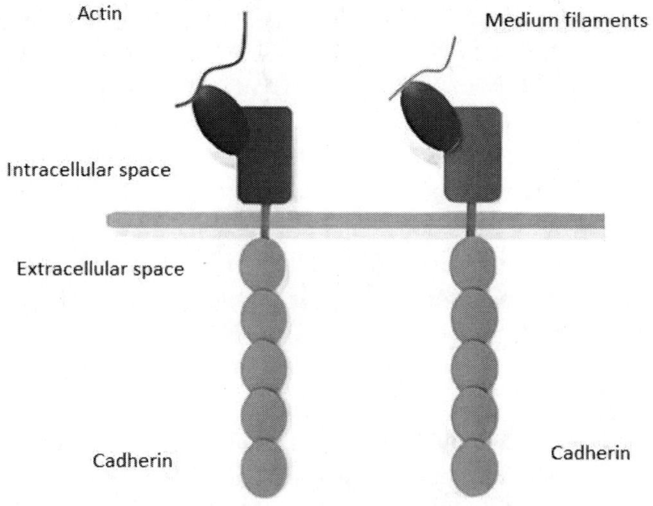


**Figure 1.** Diagram of the organization of the two types of cadhedrin (own elaboration).

Despite the fact that cadherins have been discovered for over 40 years, their development is still slow and requires significant financial outlays. Cadherins are key factors in maintaining proper tissue development. However, their incorrect expression leads to oncogenesis and metastasis. Therefore, a better understanding of cadherins is critical to cancer clinical applications, especially as therapeutic targets. Currently, cancer is the cause of premature death of one in four Europeans, and epidemiological forecasts leave no doubt that the incidence will increase with the number of new cancer cases. Cancer prevention is one of the most important challenges in the area of health protection.

We know the basics of cadherins. The E-cadherin and β-catenin complex is very important in the maintenance of epithelial morphology, and the high affinity of E-cadherin for catenin may be disrupted during oncogenesis or fibrosis. The compound is disregulated in response to inflammatory mediators, growth factors, and cytokines, suggesting that E-cadherin and β-catenin dysregulation is a promising new target for the treatment of fibrotic disorders and cancer. Higher expression of N-cadherins is associated with tumor aggressiveness, cancer metastasis, apoptosis, and angiogenesis in many human and animal neoplasms. Understanding how N-cadherin influences cell behavior will enable the development of therapies to combat its activity and prevent cancer cell growth, invasion and metastasis.

Overall, adhesion molecules are promising new targets in the treatment of cancer but could also be useful in predicting patient prognosis in human and veterinary medicine. However, more research and clinical experimentation is needed for this to happen.

## References

Cao ZQ, Wang Z, Leng P. Aberrant N-cadherin expression in cancer. *Biomed Pharmacother.* 2019;118:109320.

Corso G, Figueiredo J, De Angelis SP, Corso F, Girardi A, Pereira J, Seruca R, Bonanni B, Carneiro P, Pravettoni G, Guerini Rocco E, Veronesi P, Montagna G, Sacchini V, Gandini S. E-cadherin deregulation in breast cancer. *J Cell Mol Med.* 2020;24(11):5930-5936.

Mendonsa AM, Na TY, Gumbiner BM. E-cadherin in contact inhibition and cancer. *Oncogene.* 2018;37(35):4769-4780.

Na TY, Schecterson L, Mendonsa AM, Gumbiner BM. The functional activity of E-cadherin controls tumor cell metastasis at multiple steps. *Proc Natl Acad Sci U S A.* 2020;117(11):5931-5937.

Wong SHM, Fang CM, Chuah LH, Leong CO, Ngai SC. E-cadherin: Its dysregulation in carcinogenesis and clinical implications. *Crit Rev Oncol Hematol.* 2018;121:11-22.

# Chapter 15

# Actin

**Katarzyna Wajda**
**Dorota Bartusik-Aebisher***
**and David Aebisher**
Medical College of the University of Rzeszów, Rzeszów, Poland

## Abstract

Actin is one of the most important proteins that build the cytoskeleton and nucleoskeleton of a eukaryotic cell. It is a conserved protein, the most abundant in the cell, and it has the ability to polymerize and forms complexes with other proteins. It can undergo mutations in human organisms, usually they are missense mutations, causing systemic diseases. It correlates with a very large number of other proteins. In the human body, it is coded by 6 genes and has many important functions that enable it to function properly. Very important function performed by nuclear actin is the structure of the nucleoskeleton and the control of transcription. It was first discovered in the 19th century.

**Keywords:** actin, actomyosin, monomoeric actin, neucleoskeleton, depolymerizations, microfilamnets

Actin is one of the most important proteins that build the cytoskeleton and nucleoskeleton of a eukaryotic cell, its name comes from the Greek word "aktinos" meaning radius. It is a conserved protein, the most abundant in the

* Corresponding Author's Email: dbartusikaebisher@ur.edu.pl.

In: The Medical Biology Guide to Proteins
Editor: David Aebisher
ISBN: 979-8-88697-910-7

cell, has the ability to polymerize and forms complexes with other proteins, thanks to which it has many key functions in the cell, e.g., by interacting with myosin, it creates actomyosin and enables muscle contraction. As a protein that builds the cytoskeleton, it maintains the shape of the cell and contributes to its change, it is a scaffold for organelles located in it, is responsible for the intracellular and decellular transport of substances and the transmission of information between the inside of the cell and the environment (Blaine et al., 2020). It was first extracted from muscle cells in 1887 by the British physiologist W. D. Halliburton. It occurs not only in the cytoplasm, as it was until recently believed, but also in the cell nucleus, which was proved thanks to a pair of scientists Clark and Merriam in 1978, who in their research work isolated from the nuclear fraction of the oocytes of the plane tree a protein fulfilling similar functions and containing similar characteristics to actin derived from skeletal muscle, although it was initially thought to be the result of cytoplasm contamination of the nucleus. Nuclear actin is responsible for maintaining the proper shape of the nucleus, is involved in the transcription of, inter alia, initiates it, participates in the reconstruction of chromatin, in a complex with myosin transports fragments of active chromosomes close to euchromatin, its key functions include efficient flow and integration of signals between the cell nucleus and the cytoplasm (Blaine et al., 2020).

In the human body, actin exists in two basic forms as globular actin, monomeric G-actin and as filamentous actin-F, which occur in dynamic equilibrium determined by successive polymerizations and depolymerizations controlled by regulatory proteins - ABPs. It is composed of 375-377 amino acids and its mass is about 42.3 kDA and depends on the type of isoform and the origin of the protein it consists of. Under specific conditions depending on the external environment and the internal environment of the cell, the monomeric form of actin can convert to F-actin to form filaments. They occur in the cytosol of the cell, their thickness is 7 mm, length up to several μm, these filaments are formed into a right-handed double helix with a twist of about 166 °. Their large size prevents them from penetrating into the nucleus, because the nuclear pores are too small, therefore actin is present in the monomeric form in the nucleus. Filaments present in muscles related to specialized proteins, e.g., spectrin, emerin, titin, tropo-myosin are called thin filaments, while in non-muscle cells they are called microfilamnets (Depasquale et al., 2018).

Actin in humans occurs in the form of six structural forms, creating isoforms that differ in the number of amino acids, especially at the N-terminus, they have been distinguished due to similar functions, location and isoelectric point performed in the cell. The actin isoforms include:

- α-isoforms for skeletal and cardiac muscles
- α and γ isoforms for smooth muscles
- β and γ isoforms, the so-called cytoplasmic actins.

Monomoeric actin can move from the cytoplasm to the nucleus directly due to its small size by passive diffusion, the transport takes from 30 minutes to even several hours. The second way to move actin into the nucleus is by binding it to specialized transport proteins, i.e., cophylline. There are two fragments in the actin chain, consisting of, among others, from leucin, which are signal sites for the export of actin from the nucleus - NES. They can be inhibited by the inhibitor leptomycin B, which inhibits the transport of actin from the nucleus to the cytpolasm. After fulfilling its tasks in the nucleus, actin is removed from it, the export of actin from the cell nucleus takes place with the participation of exportin 6, which forms complexes with proteins with the NES signal, joining the porphyrin complex increases the rate of actin transport to the cytoplasm. Under stressful conditions caused by e.g., hypothermia, the intensity of actin transports to the testicle increases (Parker et al., 2020).

The distribution of actin in the nucleus depends on the function it performs and the physiological state of the cell. Nuclear actin has the ability to form complexes with proteins found in the cell nucleus, including: spectrin, α-actinin, filamine A, dystrophin, emerin, nesprin.

The functions of nuclear actin include the creation of structures responsible for the stabilization of the neucleoskeleton, which is possible thanks to the cooperation of e.g., actin, αII-spectrin, emerins. Actin is capable of forming a complex with DNAase I, which is responsible for the protection of nuclear DNA. The β isomer takes part in the initial stages of transcription, binds to RNAI, RNAII and RNAIII polymerases and controls their functioning. By affecting the enzymes modifying the work of histone proteins, actin also influences the regulation of the structure of chromosomes (Parker et al., 2020).

Actin is encoded by six genes:

- ACTA1, which encodes the α isoform
- ACTA2, ACTA3 which encode α and γ isoforms, ACTA3 encodes isoforms found in the smooth muscle of the gastrointestinal tract

- ACTC1, which encodes the α-isoform found in the heart muscle
- ACTB, ACTG1 which encode β and γ isoforms.

The ACTA1 gene is particularly vulnerable to mutations (about 220 known mutations), they are often the cause of the non-progressive Non-Maline Myopathy, the symptoms of which are respiratory muscle weakness and, consequently, problems with breathing and even with swallowing. The second most frequently mutated gene is ACTA2, which may be the cause of a family predisposition to aortic aneurysms and cerebral arteriopathy. Another mutation is the ACTC1 gene, which is responsible for the development of heart diseases such as hypertrophic or dilated cardiomyopathy. Mutations within the ACTB gene are 50% of missense mutations, disrupting the proper functioning of β-actin, which are manifested by defects such as atypical craniofacial appearance, intellectual disorders, hearing problems, heart and kidney defects and muscle atrophy characteristic of Baraitser-Winter syndrome. Mutations in this gene also contribute to abnormal bleeding. Nearly 50 mutations involve the ACTG1 gene, more than half causes deafness, while others of the remaining are responsible for Baraitser-Winter syndrome. The least common are mutations in the ACTG2 gene, the effects of which are, among others, visceral myopathy; and chronic intestinal pseudo-structure which lead to intestinal obstruction. It also plays a significant role in the development of Alzheimer's disease, as actin rods have been found to be precursors to Hirano bodies (Squire et al., 2019).

Due to the wide distribution of actin in the environment of organisms, it is a well-known protein, the subject of numerous scientific works (about 132500 search results on the pubmed.gov medical platform) and the subject of clinical and experimental research. Thanks to this research, we know that it can undergo mutations in human organisms, usually they are missense mutations, causing systemic diseases. The structure of actin is conservative, which means that it has not changed significantly in the course of evolution. It is present in the greatest amount in the cell, it is about 10%. It is composed of 375-377 amino acids and its mass is about 42.3 kDA. It consists of two main domains, each of which contains two subdomains, it is present in both eukaryotic and prokaryotic organisms. We can distinguish between cytoplasmic and nuclear actin. It correlates with a very large number of other proteins. In the human body, it is coded by 6 genes and has many important functions that enable it to function properly. One of the most important of them is to allow cells to move and maintain their correct shape, it is possible thanks to the polymerization of G-actin to F-actin and the interaction of F-actin with myosin, which in turn determines the formation of contraction. Another very important function performed by nuclear actin is the structure of the nucleoskeleton and the control of transcription. It was first discovered in the 19th century.

**Figure 1**. Actin is the main protein that enables muscle contraction.

## References

Blaine J., Dylewski J. Regulation of the Actin Cytoskeleton in Podocytes. *Cells.* 2020 Jul 16;9(7):1700.

Depasquale J. A. Actin Microridges. *Anat Rec* (Hoboken). 2018 Dec;301(12):2037-2050.

Parker F., Baboolal T. G., Peckham M. Actin Mutations and Their Role in Disease. *Int J Mol Sci.* 2020 May 10;21(9):3371.

Squire J. Special Issue: The Actin-Myosin Interaction in Muscle: Background and Overview. *Int J Mol Sci.* 2019 Nov 14;20(22):5715.

# Chapter 16

# Enkephalines

**Alicja Zając**
**David Aebisher**
**and Dorota Bartusik-Aebisher***
Medical College of the University of Rzeszów, Poland

## Abstract

Natural representatives of opioids group are endorphins, dynorphins and enkephalins. The synthetic representatives of opioids group are fentanyl, pethidine and methadone. They can be found in the neuroendocrine system, spinal cord, brain, intestinal and peripheral nervous systems, and in the endocrine part of the pancreas.

**Keywords:** enkephalines, mu-opioid receptor (MOR), tyrosine, pro-enkephalin (PENK), DNA, glucocorticoid

Opioids are a commonly defined group of natural and synthetic substances that can bind to opioid receptors. There are also semi-synthetic substances such as heroin (a product of a chemical modification of morphine), oxycodone and hydrocodone (Cullen et al., 2021). In 1975, John Hughes and Hans Kosterlitz first reported the discovery of endogenous opioids in the brain. These substances were able to inhibit the release of acetylcholine from nerves in the guinea pig's ileum. They also showed that inhibition was blocked during treatment with the antagonist naloxone (Cullen et al., 2021). Enkephalins

* Corresponding Author's Email: dbartusikaebisher@ur.edu.pl.

In: The Medical Biology Guide to Proteins
Editor: David Aebisher
ISBN: 979-8-88697-910-7

belong to the group of pentapeptides and are divided into two subgroups due to the carboxyl-terminus amino acid: leucine or methionine, which determines the division into met-enkephalin and leu-enkephalin:

- Met-enkephalin has the sequence: Tyr-Gly-Gly-Phe-Met.
- Leu-enkephalin has the sequence: Tyr-Gly-Gly-Phe-Leu.

Enkephalins are one of the three peptide systems, including β-endorphins and dynorphins. An important feature of endogenous opioid peptides is the amino acid alignment (Mongi-Bragato et al., 2018). Enkephalins act as neurotransmitters and neuromodulators throughout the nervous system and other organs of the body. The conducted studies have shown the role of met-enkephalin in the organization of tissues during the development and proliferation of cells (Mongi-Bragato et al., 2018).

Enkephalins are formed by cleavage of the precursor pro-enkephalin molecule, thanks to which met-enkephalin and leu-enkephalin are isolated in a ratio of 6: 1. These compounds are widespread in the central nervous system and in the adrenal medulla. One of the processes that enkephalins undergo is biodegradation through hydrolysis (Mongi-Bragato et al., 2021). Enkephalins are substances that are widely distributed in the body. One of the main systems is the central, peripheral and autonomic nervous systems. Leu-enkephalin and met-enkephalin have been detected in 82% of testicles and brainstem pathways of the sajmiri (monotypic subfamily of primates of the flatidae family). Increased levels of pro-enkephalin (in proportion to the severity of their condition) have been found in patients with heart failure. The short-term opioid-induced effect is characterized by a reduction in heart rate and blood pressure during the increase in contraction of the heart muscle (Mongi-Bragato et al., 2021).

Enkephalins play a very important role in the endocrine system. Studies have confirmed the direct effect of glucocorticoids on increasing the transcription of enkephalins. The role of enkephalins in modeling the time and intensity of the stress response has been demonstrated. Met-enkephalin also has an immunomodulatory effect on various types of cells, e.g., increasing the activity of CD8 + T cells or stimulating phagocytosis in macrophages (Patel et al., 2020).

Due to specific opioid receptors, enkephalins exert a physiological effect. The highest affinity of enkephalins is towards the delta-opioid receptor, the next one is for the mu-opioid and relatively low for the kappa-opioid receptor (Patel et al., 2020). Opioid receptors belong to the family of G protein-coupled receptors. By reducing the influx of $K^+$ and $Ca^{2+}$ ions, enkephalins show an inhibitory effect. Dissociation of Gα and Gβγ units is caused by transduction that begins with ligand binding. The Gα subunit directly interacts with the potassium channel causing cell hyperpolarization. This subunit causes the inhibition of adenylate cyclase, which leads to a reduction in the formation of cAMP and thus a reduction of the cAMP-dependent $Ca^{2+}$ influx. Gβγ further reduces calcium influx by binding directly to different classes of $Ca^{2+}$ channels (Tuo et al., 2020).

The extensive literature on the physiological effects of enkephalins includes, inter alia, their role in analgesia, angiogenesis, blood pressure regulation, embryonic development, feeding, hypoxia, modulation of the limbic system (emotional states), memory processes, neuroprotection, peristaltic, pancreatic secretion, wound repair, respiration control and hepatoprotective mechanisms. According to some sources, leu-enkephalin is involved in the control of gonadal function. Some of the main functions relate to analgesia, modulation of peristalsis, and regulation of the stress response.

In response to stress, the organisms release corticotropin releasing factor (CRF), which in turn stimulates the production of catecholamines while producing endogenous opioids, including enkephalins and endorphins. It has been confirmed that the endogenous opioid system is largely responsible for the intensity and duration of the stress response. An example is enkephalin that modulates the release of CRF from the medial nucleus of the hypothalamus (Tuo et al., 2020). The endogenous opioid system, due to its wide distribution throughout the nervous system and the human body, affects various disease states through its dysregulation. An example is pain dysregulation. Studies show that the availability of the mu-opioid receptor (MOR) is decreased in patients diagnosed with fibromyalgia, in addition to having elevated levels of endogenous opioids in the cerebrospinal fluid Enkephalinergic cells are scattered throughout the body. They can be found in the neuroendocrine system, spinal cord, brain, intestinal and peripheral nervous systems, and in the endocrine part of the pancreas. The enkephalin phenotype assigned to each cell type is modified and established during development by the specific environment in which the cell resides, and maintained by ongoing biosynthesis which occurs at a rate consistent with the loss of enkephalins from the cell during periods of secretion. It is now known that proenkephalin is one of three enkephalin-containing gene products, each of which gives rise to a number of physiologically active peptide products.

Enkephalins for biosynthetic expression have been studied in several types of neuroendocrine cells and tumor cell lines. Transcriptional, translational and post-translational factors may play a role in all three steps (establishment, modification, and maintenance) in regulating enkephalin expression during cell life. some of these factors are cyclic nucleotides, calcium and glucocorticoids, which pre-translative control the overall level of enkephalin biosynthesis. Posttranslational regulation of proenkephalin processing may play an important role in determining the pattern of proenkephalin-derived opioid peptides produced in a given tissue. Exogenous opioids They constitute the most effective and powerful class of agents in the treatment of acute pain associated with trauma and surgery, as well as moderate and severe cancer pain. Studying the subject of enkephalins and their associated ligands, signaling pathways and receptors that make up the opioid system is a potential for developing progressive therapies addressing a wide variety of physiological processes and organ systems. One of the more important areas of research with potential for improving treatment methods is the development of potent pain therapies that will not have the side effects associated with opionergics.

## References

Cullen J M, Cascella M. Physiology, Enkephalin. 2021 Mar 31. In: *StatPearls Treasure Island (FL)*: StatPearls Publishing; 2021 Jan–. PMID: 32491696.

Mongi-Bragato B, Avalos M P, Guzmán A S, Bollati F A, Cancela L M. Enkephalin as a Pivotal Player in Neuroadaptations Related to Psychostimulant Addiction. *Front Psychiatry. 2018* May 28;9:222.

Mongi-Bragato B, Avalos M P, Guzmán A S, García-Keller C, Bollati F A, Cancela L M. Endogenous enkephalin is necessary for cocaine-induced alteration in glutamate transmission within the nucleus accumbens. *Eur. J. Neurosci.* 2021 Mar;53(5):1441-1449.

Patel C, Meadowcroft MD, Zagon IS, McLaughlin PJ. [Met<sup>5</sup>]-enkephalin preserves diffusion metrics in EAE mice. *Brain Res Bull.* 2020 Dec;165:246-252.

Tuo Y, Tian C, Lu L, Xiang M. The paradoxical role of methionine encephalin in tumor responses. *Eur. J. Pharmacol.* 2020 Sep 5;882:173253.

## Chapter 17

# MAP17 in Laryngeal Cancer

**Lidia Bieniasz**
**David Aebisher**
**and Dorota Bartusik-Aebisher***
Medical College of the University of Rzeszów, Rzeszów, Poland

## Abstract

MAP17 is a small (17 kDa), non-glycosylated membrane protein located in the plasma membrane and the Golgi apparatus. The physiological role of this protein in the proximal tubules is not well understood, however, MAP17 stimulates SGLT transporters, increasing the specific Na-dependent transport of mannose and glucose in oocytes and human tumor cells. MAP17 is overexpressed, mainly through mRNA amplification, in various human cancers. Through the enhanced tumorigenic properties induced by MAP17, they are associated with an increase in ROS, since MAP17 significantly alters the mRNA levels of genes involved in oxidative stress and increases endogenous ROS, and antioxidant treatment of MAP17-expressing cells reduces their cell carcinogenic properties. Generalized overexpression of MAP17 in human cancers indicates that MAP17 may be a good marker of neoplasm, especially malignant progression.

**Keywords:** MAP17, neoplastic cells, prostate cancer, breast cancer, sodium-dependent glucose transporter 1 (SGLT1)

---

* Corresponding Author's Email: dbartusikaebisher@ur.edu.pl.

In: The Medical Biology Guide to Proteins
Editor: David Aebisher
ISBN: 979-8-88697-910-7

MAP17 is a small (17 kDa), non-glycosylated membrane protein located in the plasma membrane and the Golgi apparatus. The physiological role of this protein in the proximal tubules is not well understood, however, MAP17 stimulates SGLT transporters, increasing the specific Na-dependent transport of mannose and glucose in oocytes and human tumor cells (Blasco et al., 2003).

Neoplastic cells in which overexpression of MAP17 has been observed are characterized by increased neoplastic phenotypes with increased proliferative capacity in the presence or absence of contact inhibition, decreased sensitivity to apoptosis and increased migration. Cell behavior is closely related to MAP17 induction and is associated with an increase in ROS production, and treatment of MAP17-expressing cells with antioxidants reduces carcinogenic properties. MAP17 is overexpressed, mainly through increased mRNA levels, in various cancers. Expression of MAP17 increases ROS through SGLT1 in neoplastic cells and therefore MAP17 and SGLT1 may be markers of high oxidative stress tumors and thus further increase of ROS may raise levels beyond the apoptotic threshold. Thus, MAP17 may be a marker of the activity of therapies in which oxidative stress plays a key role in the response (Guijarro et al., 2007). The MAP17 protein is overexpressed in a large percentage of the analyzed tumors and is significantly correlated with the stage of ovarian, prostate and breast cancer. In tumors such as the nipple, ovary, colon, stomach, cervix, and thyroid gland, the overexpression rate in tumor samples is greater than 70%, while in the lung, uterus, and rectum it is about 50%. Literature data suggest that the markers MAP17 and SGLT1 can be used to identify patients who are likely to show a better response to treatments that increase oxidative stress in other types of cancer. In practice, cervical cancer cells overexpressing MAP17 are more sensitive to factors capable of inducing oxidative stress (Guijarro et al., 2007).

Neoplastic cells overexpressing MAP17 are characterized by increased proliferative capacity. The expression of MAP17 is associated with an increase in ROS mediated by SGLT, which acts as a second messenger to enhance tumorigenesis. Even a small increase in ROS activates signaling cascades that enhance neoplastic processes, where further increases in ROS lead to a potentially toxic cellular environment and programmed cell death (Guijarro et al., 2007). Tumors expressing high levels of ROS producing the proteins MAP17 and SGLT1 may benefit from treatments such as cisplatin or radiation therapy which increase oxidative stress and may sensitize them to cell death. In this situation, the analysis of laryngeal cancer has a significant relationship between the high expression of the MAP17 protein, suggesting that the expression of MAP17 is an independent biomarker of survival. The available data confirm that MAP17 in combination with SGLT1 is a good prognostic marker of survival in patients with laryngeal cancer treated with radiotherapy and chemotherapy. Consequently, MAP17 can predict which patients may have better survival outcomes. Further prospective and controlled studies are needed to confirm our results and validate MAP17 as a new biomarker for clinical use in laryngeal cancer.

MAP17 is overexpressed, mainly through mRNA amplification, in various human cancers. Based on the literature data, the immunohistochemical analysis of MAP17 during tumor progression showed that overexpression strongly correlates with tumor progression in prostate, breast and ovarian cancer. Generalized overexpression of MAP17 in human cancers indicates that MAP17 may be a good marker of neoplasm, especially malignant progression. Through the enhanced tumorigenic properties induced by MAP17, they are associated with an increase in ROS, since MAP17 significantly alters the mRNA levels of genes involved in oxidative stress and increases endogenous ROS, and antioxidant treatment of MAP17-expressing cells reduces their cell carcinogenic properties. Therefore, ROS is generated in a MAP17-dependent manner as an intracellular signal and induces a genetic program associated with growth. On the other hand, sodium-dependent glucose transporter 1 (SGLT1) appears to mediate intracellular glucose uptake by MAP17 and increase ROS. Together, + that MAP17 dependent tumorigenic properties are dependent on ROS activation via glucose increase via SGLT1 membrane transport (de Miguel-Luken et al., 2015).

## References

Blasco T., Aramayona J. J., Alcalde A. I., Catalán J., Sarasa M., Sorribas V. Rat kidney MAP17 induces cotransport of Na-mannose and Na-glucose in Xenopus laevis oocytes. *Am J Physiol Renal Physiol.* 2003; 285(4):F799-810.

de Miguel-Luken M. J., Chaves-Conde M., de Miguel-Luken V., Muñoz-Galván S., López-Guerra J. L., Mateos J. C., Pachón J., Chinchón D., Suarez V., Carnero A. MAP17 (PDZKIP1) as a novel prognostic biomarker for laryngeal cancer. *Oncotarget*. 2015 May 20;6(14):12625-36.

Guijarro M. V., Leal J. F., Blanco-Aparicio C., Alonso S., Fominaya J., Lleonart M., Castellvi J., Ramon y Cajal S., Carnero A. MAP17 enhances the malignant behavior of tumor cells through ROS increase. *Carcinogenesis*. 2007;28(10):2096-104.

Guijarro M. V., Leal J. F., Fominaya J., Blanco-Aparicio C., Alonso S., Lleonart M., Castellvi J., Ruiz L., Ramon Y. Cajal S., Carnero A. MAP17 overexpression is a common characteristic of carcinomas. *Carcinogenesis.* 2007;28(8):1646-52.

Guijarro M. V., Link W., Rosado A., Leal J. F., Carnero A. MAP17 inhibits Myc-induced apoptosis through PI3K/AKT pathway activation. *Carcinogenesis.* 2007;28(12):2443-50.

## Chapter 18

# Erythropoietin Overexpression in Head and Neck Tumors

**Lidia Bieniasz**
**David Aebisher**
**and Dorota Bartusik-Aebisher***
Medical College of the University of Rzeszów, Poland

## Abstract

Erythropoietin (EPO) is a glycoprotein hormone that is naturally produced by the peritubular cells of the kidney to stimulate the production of red blood cells. Literature results confirmed the presence of Epo and EpoR in malignant tumors of the larynx and showed a correlation between Epo expression and survival. Hypoxia is the primary stimulus for the production of erythropoietin, which acts through the erythropoietin receptor. As a result of hypoxia within the tumor, the transcription factor HIF, and in particular HIF-1α, induces the production of the glycoprotein hormone erythropoietin (EPO) in kidney and liver cells. The concentration of EPO and EPO-R in cancers of the oral cavity, pharynx and larynx is significantly increased compared to healthy tissues. The accurate diagnosis and subsequent prediction of the course of a malignant larynx disease remains a challenge.

**Keywords:** erythropoietin, head and neck cancer, laryngeal cancer

---

* Corresponding Author's Email: dbartusikaebisher@ur.edu.pl.

In: The Medical Biology Guide to Proteins
Editor: David Aebisher
ISBN: 979-8-88697-910-7

Erythropoietin (EPO) is a glycoprotein hormone that is naturally produced by the peritubular cells of the kidney to stimulate the production of red blood cells. Peritubular cells of the renal cortex produce most of the EPO in the human body. Erythropoietin is hormonally secreted and binds to receptors on the surface of immature red blood cells, influencing their differentiation and proliferation. As a result of these processes, red blood cells are formed, thus preventing the effects of hypoxia of the tumor tissue. When EPO-R is combined with erythropoietin, a signaling pathway that influences angiogenesis and inhibits cell apoptosis is activated. EPO secretion is regulated primarily by the oxygen level in the blood. Hypoxia, is the state of hypoxia, becomes the main stimulus for EPO secretion (Semenza et al. 2009).

Erythropoietin is a lunar protein because it can act as a hormone, cytokine, and growth factor. Its main function is to regulate the production of red blood cells in the bone marrow. However, EPO and its EPOR receptors are also expressed in non-hematopoietic tissues such as the endothelium, where they play a protective role. In addition, it is known that the EPO-EPOR pathway contributes to the formation of new blood vessels in tumor angiogenesis, but its mechanism is not fully understood. In this chapter, after a brief introduction to neoplastic angiogenesis and a description of classical and non-classical proangiogenic factors, we have reviewed the role of EPO in neoplastic angiogenesis, focusing on the different activation mechanisms conducive to tumor growth and progression. Careful characterization of EPO variants and their downstream pathways will allow the development of specific strategies to inhibit only the blocking of EPOR expressed by tumor cells without inducing signal transduction in hematopoietic cells in order to avoid side effects. EPO, VEGF and their receptors are hypoxia-sensitive genes regulated by the transcription factors HIF-1, HIF-2 and HIF-3. Mechanism-wise, under hypoxic conditions, the regulatory subunit of HIF-α becomes stable and dimerizes with the HIF-β subunit constitutively expressed in the nucleus, forming the HIF heterodimer The HIF heterodimer is reconnected to the hypoxic response enhancing region of EPO and other oxygen-sensitive genes (Seibold et al. 2013).

Based on literature data, the presence of Epo and EpoR in malignant laryngeal tumors has been confirmed so far and we have demonstrated a correlation between Epo expression and survival. More research is needed to further define the role of Epo and EpoR in the treatment of patients with laryngeal cancer (Annese et al. 2019).

The next step will be to investigate whether the laryngeal cancer cells express erythropoietin (Epo) and the erythropoietin receptor (EpoR) and their possible relationship to the clinical and pathological features of the tumor, taking into account the fact that laryngeal cancer cells express Epo and EpoR and what is the their possible relationship with the clinical and pathological features of the tumor (Kodetthoor et al. 2006).

The hallmark of tumor growth and expansion is that the local vascular system cannot supply enough oxygen and nutrients for the therapeutic division of cancer cells. The resulting hypoxia can alter neoplastic cells leading to cell arrest or apoptosis, or leading to an improvement in tissue oxygen response. It reduces apoptotic potential, providing growth benefits to cells with genetic alterations that impair apoptosis. It is now well known that EPO and EPOR are associated not only with the erythropoiesis of the bone marrow itself, as it has been found that they are activated in most other normal cells and neoplastic cells. In these cells, the EPO-EPOR pathway enhances cell proliferation, inhibits cell apoptosis, and promotes physiological and pathological mechanisms of angiogenesis (Mohyeldin et al. 2009).

Despite the many available diagnostic procedures, accurate diagnosis and subsequent prediction of the course of a malignant larynx disease remain a challenge. The current challenge is to investigate the role of erythropoietin and the erythropoietin receptor in malignant laryngeal tissue as new markers that can predict the course of the disease. Literature results confirmed the presence of Epo and EpoR in malignant tumors of the larynx and showed a correlation between Epo expression and survival. Hypoxia is the primary stimulus for the production of erythropoietin, which acts through the erythropoietin receptor. Epo was once thought to act exclusively on erythroid precursor cells and was reported to play a role in promoting growth, inhibiting apoptosis, and inducing erythroid differentiation. As a result of hypoxia within the tumor, the transcription factor HIF, and in particular HIF-1$\alpha$, induces the production of the glycoprotein hormone erythropoietin (EPO) in kidney and liver cells. It has been shown that in the biopsy material of cancers of the oral cavity, pharynx and larynx, the concentration of EPO and EPO-R is significantly increased compared to healthy tissues. Another study showed that administration of recombinant erythropoietin to treat anemia resulting from radiotherapy may adversely affect the prognosis of patients with head and neck tumors if the cancer cells express EPO-R.

## References

Annese, T., Tamma, R., Ruggieri, S., and Ribatti, D. (2019). Erythropoietin in tumor angiogenesis. *Exp Cell Res.*, 374(2), 266-273.

Mohyeldin, A., Lu, H., Dalgard, C., Lai, S. Y., Cohen, N., Acs, G., and Verma, A. (2005). Erythropoietin signaling promotes invasiveness of human head and neck squamous cell carcinoma. *Neoplasia.*, 7(5), 537-43

Seibold, N. D., Schild, S. E., Gebhard, M. P., Noack, F., Schröder, U., and Rades, D. (2013 Jul). Prognosis of patients with locally advanced squamous cell carcinoma of the head and neck. Impact of tumor cell expression of EPO and EPO-R. *Strahlenther Onkol.*, 189(7), 559-65.

Semenza, G. L. (2009 Apr). Regulation of oxygen homeostasis by hypoxia-inducible factor 1. *Physiology (Bethesda).*, 24, 97-106.

Udupa, K. B. (2006). Functional significance of erythropoietin receptor on tumor cells. *World J Gastroenterol*, 12(46), 7460-7462.

## Chapter 19

# PD-L1 in Head and Neck Neoplasms

**Lidia Bieniasz**
**David Aebisher**
**and Dorota Bartusik-Aebisher***
Medical College of the University of Rzeszów, Rzeszów, Poland

## Abstract

PD-L1 is a validated biomarker used to guide treatment choice in clinical practice. Over the past 10 years, cancer immunotherapy has made significant advances in many cancers and is gradually being applied in clinical oncology care, including Programmed Cell Death Protein-1 (PD-1)/Programmed Cell Death Ligand 1 (PD-L1 Access). Compared to traditional therapies, the emerging PD-1/PD-L1 blocking immunotherapy shows more satisfactory therapeutic effects and less toxicity in patients with advanced squamous cell carcinoma of the head and neck. Expression of PD-L1 on cells of the immune system in a tumor biopsy before treatment indicates a previously induced adaptive anti-tumor immune response and is associated with improved treatment outcomes. Understanding the complex mechanism behind PD-L1 presentation in TME may enable therapeutic approaches to regulate the expression of this immunosuppressive ligand to enhance the PD-1 blockade effect.

**Keywords:** PD-L1; head and neck cancer; expression of PD-L1

* Corresponding Author's Email: dbartusikaebisher@ur.edu.pl.

In: The Medical Biology Guide to Proteins
Editor: David Aebisher
ISBN: 979-8-88697-910-7

The PD-1 checkpoint receptor expressed on activated T cells promotes immunosuppression after interaction with its ligands PD-L1 and PD-L2, which are found in cancer cells and cells that attack the immune system. Expression of PD-L1 on cells of the immune system in a tumor biopsy before treatment indicates a previously induced adaptive anti-tumor immune response and is associated with improved treatment outcomes (Brahmer et al., 2012). Therefore, blocking the interaction of PD-1 / PD-L1 with anti-PD-1 or anti-PD-L1 mAb may promote immune reactivation, thus causing persistent resistance in some patients with various solid tumors. Achievements so far show that the total positive (CPS) ratio, the ratio of the number of PD-L1 positive cells, including tumors, lymphocytes and macrophages, to the total number of neoplastic cells, provides a more effective result than the tumor index ratio. TPS) measures PD-L1 expression in neoplastic cells. Preclinical data indicate that squamous cell carcinoma of the head and neck (HNSCC) is a serious immunosuppressive disease characterized by inappropriate secretion of proinflammatory cytokines and dysfunction of immune effector cells. Although only PD-L1 is now widely used as a predictive biomarker of the immune checkpoint suppression response in HNSCC, there is still much research ongoing to identify new biomarkers (Gavrielatou et al., 2020).

Compared to inflammation-induced PD-L1 expression, oncogene-induced PD-L1 expression is a different unit in histopathology and biology (Qiao et al., 2020). Although the latter focuses on IFNγ-mediated immune attack sites, oncogene-mediated PD-L1 expression is constitutive and diffuse. In glioblastoma multiforme, activation of Akt and Ras coupled with loss of PTEN leads to the highest level of PD-L1 expression in IHC, while reducing in vitro T cell mortality (Sharpe et al., 2007). Inhibition of PD-1 in this chimeric EGFR-based mouse model resulted in tumor regression and improved survival. In general, more and more signaling pathways are associated with the regulation of PD-L1, supporting the search for combination therapies including TKI and immunotherapy (Parsa et al., 2007).

Currently, only PD-L1 is a validated biomarker used to guide treatment choice in clinical practice. Research should focus on improving patient selection through the implementation of PD-L1 and the identification of new prognostic biomarkers. Over the past 10 years, cancer immunotherapy has made significant advances in many cancers and is gradually being applied in clinical oncology care, including Programmed Cell Death Protein-1 (PD-1) / Programmed Cell Death Ligand 1 (PD-L1 Access) is one of the the most attractive goals. Compared to traditional therapies, the emerging PD-1 / PD-L1 blocking immunotherapy shows more satisfactory therapeutic effects and less toxicity in patients with advanced squamous cell carcinoma of the head and neck.

Many cytokines found in immunoreactive TME can induce PD-L1 expression on tumor and/or immune cells through various signaling mechanisms. In this study, no factors influencing constitutive PD-L1 expression were found. Understanding the complex mechanism behind PD-L1 presentation in TME may enable therapeutic approaches to regulate the expression of this immunosuppressive ligand to enhance the PD-1 blockade effect.

## References

Brahmer J R, Tykodi S S, Chow L Q, Hwu W J, Topalian S L, Hwu P, Drake C G, Camacho L H, Kauh J, Odunsi K, Pitot H C, Hamid O, Bhatia S, Martins R, Eaton K, Chen S, Salay T M, Alaparthy S, Grosso J F, Korman A J, Parker S M, Agrawal S, Goldberg S M, Pardoll D M, Gupta A, Wigginton J M. Safety and activity of anti-PD-L1 antibody in patients with advanced cancer *N. Engl. J. Med.*, 366 (2012), pp. 2455-2465.

Gavrielatou N, Doumas S, Economopoulou P, Foukas P G, Psyrri A. Biomarkers for immunotherapy response in head and neck cancer. *Cancer Treat. Rev.* 2020 Mar;84:101977.

Parsa A T, Waldron J S, Panner A, Crane C A, Parney I F, Barry J J, Cachola K E, Murray J C, Tihan T, Jensen M C, Mischel P S, Stokoe D, Pieper R O. Loss of tumor suppressor PTEN function increases B7-H1 expression and immunoresistance in glioma. *Nat. Med.* 2007;13:84–8.

Qiao X W, Jiang J, Pang X, Huang M C, Tang Y J, Liang X H, Tang Y L. The Evolving Landscape of PD-1/PD-L1 Pathway in Head and Neck Cancer. *Front. Immunol.* 2020;11:1721.

Sharpe A H, Wherry E J, Ahmed R, Freeman G J. The function of programmed cell death 1 and its ligands in regulating autoimmunity and infection. *Nat. Immunol.* 2007;8:239–45.

# Chapter 20

# Serotonin Receptors

**Julia Inglot**
**Dorota Bartusik-Aebisher***
**and David Aebisher**
Medical College of the University of Rzeszów, Poland

## Abstract

Although serotonin is produced by a very small number of neurons, it has countless functions in the human body. By affecting the digestive, circulatory, endocrine, urogenital and central nervous systems, it controls such processes as respiration, metabolism, digestion, heart function, vascular contractility, blood homeostasis, micturition, reproduction and behavioral processes. The possible use of excessive or reduced stimulation of serotonin receptors in the treatment of many diseases is at the stage of research. Due to differences in structure, function, action and ligands, 5-HT receptors have been divided into 7 families and at least 15 variants. Most of them belong to the receptors associated with the G protein, 5-HT3 belongs to the ionic receptors.

**Keywords:** serotonin, serotonin receptor, 5-HT, 5-hydroxytryptamine

Research on serotonin (5-hydroxytryptamine, 5-HT), which dates back to the 1940s, provides more and more new information about its effects on most organs in the human body. In the 1950s, evidence of the action of 5-hydroxytryptamine as a neurotransmitter in the central nervous system of

* Corresponding Author's Email: dbartusikaebisher@ur.edu.pl.

In: The Medical Biology Guide to Proteins
Editor: David Aebisher
ISBN: 979-8-88697-910-7

animals emerged, and in the following decades more variants were identified. Currently, there are at least 15 5HT receptors, grouped into 7 families, but the indication of the function of a given subpopulation is difficult due to the lack of selective agents. In addition, one receptor has several functions, causing a drug that is selective to act to be able to affect organs other than its target organ as well. 1,2,4 Given that serotonin is produced by less than 1 neuron per million, the fact that it produces a large proportion of the body's activity may be surprising (Ikarashi et al. 2018).

Work is underway on the use of positron emission tomography (PET) and single photon emission tomography (SPECT) imaging to compare the functioning of the 5-HT system in vivo depending on age. The first studies showed that aging in the body negatively affects serotonin receptors, especially 5-HT1A and 5-HT2A, and as serotonin regulates emotional, physiological and cognitive processes, these changes correlate with changes in the body. It explains what may underlie changes related to the aging of the human body.

Serotonin is a monoamine, identified by Vialli and Erspamer in 1937. About 95% of the body's total serotonin is released into the intestines by pheochromocytomas - from there it was isolated for the first time, and its original name was enteramine. Its chemical structure is similar to tryptamine, dimethyltryptamine, diethytryptamine, melatonin and bufotein, which are indole alkylamines. 5-HT biosynthesis begins with the hydroxylation of L-tryptophan. Subsequently, with the participation of tryptophan hydroxylase, L-tryptophan is converted to hydroxytryptophan, and thanks to 5-hydroxy-tryptophan decarboxylase, it is decarboxylated to 5-hydroxytryptophan.

With the exception of 5-HT3, belonging to the ionotropic receptors, serotonin receptors are associated with the G protein. The 5-HT1 and 5-HT5 families, having an inhibitory effect, bind to the Gi/G0 protein, which reduces the cellular cAMP level. The stimulants 5-HT4, 5-HT6 and 5-HT7 bind to the Gs protein, causing an increase in the level of cellular cAMP (cyclic adenosine monophosphate). On the other hand, 5-HT2, also with a stimulating effect, by binding with the Gq11 protein, increases the cellular level of IP3 (inositol triphosphate) and DAG (diacylglycerol). Activation of the 5-HT3 receptor results in depolarization of the cell membrane, which occurs with the help of ion channels for $Na^+$ and $K^+$.

The G protein is a heterotrimeric complex composed of the Gα, Gβ and Gγ subunits, the latter two together forming the βγ complex. The subunits are located on the inner side of the cell membrane and induce activation of the G protein-coupled receptor (GPCR). Binding of the ligand to the GPCR domain outside the cell triggers a cascade of processes, consequently changing the function of a given cell by stimulating or inhibiting the intracellular pathway. Several G proteins can bind to the same serotonin receptor but induce a different signaling pathway.

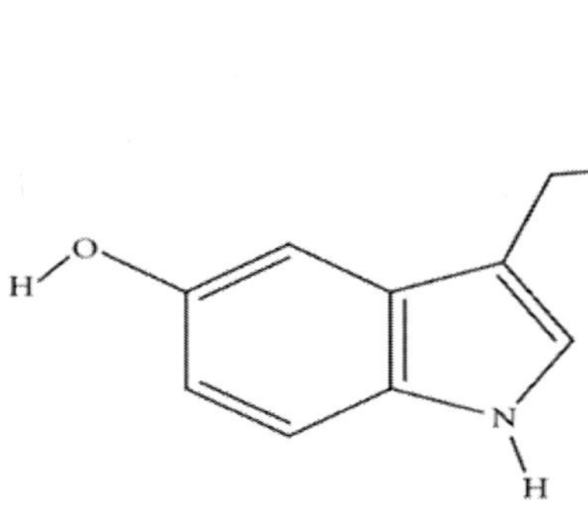


**Figure 1.** Serotonin (5-HT). Modified image from NCBI PubChem Substance Database CID 5202.

Serotonin is found mainly in nerve endings, while it is also found in other parts of the body. Its action reaches almost every organ of the human body. 5-HT regulates most behavioral processes, including mood, anxiety, perception, reward, anger, aggression, sexuality, and appetite. It also plays an important role in the learning process, being responsible for attention as well as short and long-term memory. By influencing the circulatory system, it controls, among others vascular resistance, blood pressure, haemostasis, platelet function and vasodilatation, as well as electrical conduction of the heart, valve closure and myocardial remodeling after infarction. By acting on the breathing centers in the brainstem and the pulmonary vascular system, serotonin helps monitor breathing, rhythm, and respiratory drive.

By affecting the endocrine system and metabolism, it regulates the energy balance, including glucose metabolism, through the 5-HT2C receptor, and with the participation of 5-HT2C and 5HT1B, it stimulates the innervation of brown adipose tissue. It controls the hypothalamic-pituitary-adrenal axis, temperature, as well as the development and regeneration of metabolic and endocrine organs (Karmakar et al. 2021). 5-HT regulates digestion at many stages. When the tongue's taste buds are activated by food, serotonin is released into the sensory afferent nerves, transmitting taste information to the CNS. After food enters the gastrointestinal tract, it stimulates peristaltic waves and the secretion of pancreatic enzymes. Serotonin affects the genitourinary system both centrally and peripherally. Its main action is to delay ejaculation and orgasm, inhibit voiding, uterine contractions, stimulating sperm transport to the fallopian tube and inducing the expression of collagenase in the uterus after delivery, facilitating its involution during the puerperium. Increased serotonin levels are observed during pregnancy, and its amount correlates with the severity of pre-eclampsia (Oliver et al. 2021).

Disturbances in the functioning of the serotin receptor may be the basis of depression. This mainly concerns 5-HT1A, 5-HT1B, 5-HT4 and 5-HT7 receptors, whose reduced stimulation significantly reduces the mood. However, the causes of this disease are not yet fully understood, which is confirmed by the fact that about 1/3 of patients do not respond to treatment with a selective serotonin reuptake inhibitor. In addition to depression, it is believed that 5-HT-dependent signaling abnormalities may lead to other psychiatric disorders, i.e., schizophrenia, Alzheimer's disease or autism (Lee et al. 2021).

Excessive stimulation of serotonergic receptors may lead to atrial fibrillation, remodeling of the ventricles in their failure, enlargement of the area and fibrosis of the valve leaflet, and consequently deterioration of their closure. Thus, 5-HT2A receptor antagonists may be useful in the treatment of vasospastic angina and ischemic heart disease, and 5-HT3 in the treatment of pain after myocardial infarction. The absence of the 5-HT2B receptor has been associated with heart defects or dilated cardiomyopathies in adults (Lee et al. 2021).

Serotonin supports the respiratory process, while in the case of hypoxia, its excess increases the signaling of the mitogenic 5-HT2B receptor on the endothelial cells of the pulmonary artery, increasing vascular resistance, which results in pulmonary hypertension and, secondarily, right ventricular hypertrophy. Accordingly, 5-HT2B antagonists find use in the treatment of early pulmonary arterial hypertension. Research shows that activation of the 5-HT4 receptor has a stimulating effect on the respiratory system, which suggests the possibility of using agonists of this receptor as a means of inhibiting respiratory depression. Moreover, serotonergic abnormalities have been found in approximately half of the babies who died from sudden infant death. A significantly greater number of serotonin medullary neurons, which are central chemoreceptors, but with lower expression of the 5HT1A receptor and transporter, were observed in them (Oliver et al. 2021).

Too little stimulation of 5-HT2C receptors disturbs the regulation of energy balance and glucose metabolism, leading to obesity and diabetes, therefore antagonists can be used to treat these disorders. By controlling the hypothalamic-pituitary-adrenal axis, serotonin deficiency may lead to hypothermia or a disturbed response to stress, and additionally, by regulating epithelial connections and the work of secretory glands, it plays an important role in their development, regeneration and renewal. Impaired serotonin receptor signaling is associated with functional gut disorders, e.g., irritable bowel syndrome, and drugs that affect 5-HT3 and 5-HT4 receptors can therefore be used to treat these conditions. Studies have shown that 5-HT1B and 5-HT2B receptors were present in more than 30% of patients with hepatocellular carcinoma, and the use of antagonists of these receptors reduced disease progression. Excessive stimulation of serotonin receptors in the digestive system, especially by activation of 5-HT3 receptors in the vagus nerves innervating emetic centers in the brainstem, causes vomiting. Thus, 5-HT3 antagonists are effective anti-emetic agents. Altered serotonergic modulation may result in mood disorders and increased pain perception. The excess of serotonin can significantly inhibit ejaculation and micturition, therefore selective serotonin reuptake inhibitors have found application in the treatment of premature ejaculation and stress urinary incontinence. Although elevated serotonin levels are physiological during pregnancy, research has shown that this amount correlates with the severity of pre-eclampsia. Therefore, blocking serotonin receptors seems to be a good way to treat this pathology (Prasad et al. 2019).

Serotonin receptor ligands can be categorized into agonists and antagonists that are distinct for each receptor family (Table 1).

Serotonin, known primarily as a neurotransmitter, in addition to the central nervous system also affects the circulatory, digestive, endocrine and genitourinary systems. Its main task is to regulate physiological, emotional and cognitive processes. It often takes place in many stages, through opposing mechanisms - the strengthening or inhibition of nerve impulses, which allows the organism to adapt to the changing conditions of the external environment. Properties of individual receptor varieties, i.e., structure, mode of action and compatible ligands determine different functions. Disturbances in their action are responsible for a number of diseases but are also a therapeutic possibility for others. Where insufficient or overstimulation of the serotonergic receptor underlies the pathomechanism, it may be curative in other conditions. Hundreds of studies have been carried out on the 5-HT receptor, but there are still many uncertainties that, if further explored, could provide solutions to important problems.

**Table 1.** Subpopulations of 5-HT receptors families

| Receptor | Agonists | Antagonists |
|---|---|---|
| 5-HT1A | buspiron, dihydroergotamine, eltoprazine, ergotamine, flesinoxan, flibanserin, gepirone, ipsapirone, methysergide, quetiapine, tandospirone, urapidil, yohimbine, ziprasidone | spiperone, alprenolol, asenapine, cyanopindolol, iodocyanopindolol, lecozotan, methiothepin, oxprenolol, pindolol, propanolol |
| 5-HT1B | dihydroergotamine, eletriptan, eltoprazine, ergotamine, methysergide, sumatriptan, zolmitriptan | yohimbine, alprenolol, asenapine, cyanopindolol, iodocyanopindolol, isamoltane, metergoline, methiothepin, oxprenolol, pindolol, propanolol |
| 5-HT1D | sumatriptan, almotriptan, dihydroergotamine, eletriptan, ergotamine, frovatriptan, methysergide, naratriptan, rizatriptan, yohimbine, zolmitriptan | ketanserin, metergoline, methiothepin, rRauwolscine, ritanserin |
| 5-HT1E | eletriptan, methysergide, tryptamine | methiothepin |
| 5-HT1F | eletriptan, naratriptan, sumatriptan | methiothepin |
| 5-HT2A | bufotenin, ergonovine, lisuride, LSD (in CNS), mescaline, myristicin, psilocin, psilocybin, yohimbine | aripiprazole, clozapine, cyproheptadine, eplivanserin, etoperidone, iloperidone, ketanserin, methysergide, mirtazapine, nefazodone, olanzapine, quetiapine, risperidone, ritanserin, trazodone, ziprasidone |
| 5-HT2B | α-metyl-5-HT, fenfluramine, LSD (in CNS), norfenfluramine | agomelatine, asenapine, ketanserin, LSD (PNS), methysergide, ritanserin, tegaserod, yohimbine |
| 5-HT2C | α-metyl-5-HT, aripiprazole, ergonovine, lorcaserin, LSD (in CNS) | agomelatine, asenapine, clozapine, cyproheptadine, eltoprazine, etoperidone, fluoxetine, ketanserin, lisuride, LSD (in PNS), methysergide, mianserin, mirtazapine, nefazodone, olanzapine, risperidone, ritanserin, trazodone, ziprasidone |
| 5-HT3 (5-HT3A, 5-HT3B) | α-metyl-5-HT, quipazine | alosetron, clozapine, dolasetron, granisetron, memantine, metoclopramide, mianserin, mirtazapine, olanzapine, ondansetron, quetiapine, tropisetron |
| 5-HT4 (5-HT4A-H) | cisapride, metoclopramide, mosapride, prucalopride, renzapride, tegaserod,zacopride | L-lysine, piboserod |
| 5-HT5 (only 5-HT5A receptor in humans) | ergotamine, valerenic acid | asenapine, dimebolin, methiothepin, ritanserin |
| 5-HT6 | EMD-386.088, EMDT | aripiprazole, asenapine, clozapine, dimebolin, iloperidone, olanzapine |
| 5-HT7 | 5-carboxytryptamin, LSD | aripiprazole, asenapine, clozapine, iloperidone, ketanserin, metiotepín, olanzapine, ritanserin |

CNS: central nervous system; PNS: peripheral nervous system; LSD: lysergic acid diethylamide (lysergide); EMDT: 2-ethyl-5-methoxy- N,N-dimethyltryptamine; GI: gastrointestinal. Sources: [1], Berger et al. 2009, Nichols and Nichols 2008, Nelson 2004.

## References

Ikarashi Y, Sekiguchi K, Mizoguchi K. Serotonin Receptor Binding Characteristics of Geissoschizine Methyl Ether, an Indole Alkaloid in Uncaria Hook. *Curr Med Chem.* 2018;25(9):1036-1045.

Karmakar S, Lal G. Role of serotonin receptor signaling in cancer cells and anti-tumor immunity. *Theranostics.* 2021 Mar 11;11(11):5296-5312.

Lee J, Avramets D, Jeon B, Choo H. Modulation of Serotonin Receptors in Neurodevelopmental Disorders: Focus on 5-HT7 Receptor. *Molecules.* 2021 Jun 2;26(11):3348.

Oliver BL, Pahua AE, Hitchcock K, Amodeo DA. Serotonin 6 receptor modulation reduces locomotor activity in C57BL/6J mice. *Brain Res.* 2021 Apr 15;1757:147313.

Prasad S, Ponimaskin E, Zeug A. Serotonin receptor oligomerization regulates cAMP-based signaling. *J Cell Sci.* 2019 Aug 23;132(16):jcs230334.

# Chapter 21

# GABA Receptors: Structure, Functioning, Ligands and Selected Disorders

**Maksymilian Kłosowicz**
**Dorota Bartusik-Aebisher***
**and David Aebisher**
Medical College of the University of Rzeszów, Poland

## Abstract

GABA receptors are a group of membrane receptors that are widely distributed not only within the central nervous system but also in other organs. Based on their structural structure, GABA receptors have been divided into 4 classes: A, B, C and F. Each of them is characterized by a different structure, which determines its pharmacokinetic properties, the degree of binding with various ligands and the functions of the receptor. The GABAergic system has a number of functions. The influence of GABA receptors begins in fetal life, where they ensure the correct process of shaping neural networks. Later, they watch over such processes as learning, memory, and the proper shaping of emotions and behavior. They play an important role in the pathomechanism of many CNS diseases (Ghit et al. 2021).

**Keywords:** γ-aminobutyric acid, GABA receptors, CNS, Alzheimer's disease, benzodiazepines

---

* Corresponding Author's Email: dbartusikaebisher@ur.edu.pl.

In: The Medical Biology Guide to Proteins
Editor: David Aebisher
ISBN: 979-8-88697-910-7

γ-Aminobutyric acid (GABA) receptors are a group of membrane receptors that are found not only in the central nervous system (CNS) but also in other organs in the body. Based on the structure of transmembrane domains, GABA receptors have been divided into 4 classes: A, B, C and F. GABAB receptors belong to metabotropic receptors, their activation leads to the opening of potassium channels and hyperpolarization of the cell membrane. In contrast, GABAA and GABAC are examples of ionotropic receptors. Activation of this type of receptor leads to the transport of chloride ions inside the cell, which also causes hyperpolarization of the cell membrane and inhibition in the cell. The following publication summarizes the current knowledge about the structure, location and function of the GABAergic receptor and its specific isoforms and ligands binding to the receptor and the pharmacological action they exert. The last subsection deals with topics related to the pathophysiology of the GABA system in selected patients' disease entities (Ghit et al. 2021).

GABAA receptors belong to ligand-gated ion channels, stimulation of which inhibits synaptic conduction. They are made up of five protein subunits, each of which has approximately 450 amino acid residues. Within a single subunit, we can distinguish the N-terminal extracellular domain with a Cys loop, 4 transmembrane sequences (M1-M4), a large intracellular loop (between M3 and M4) and the extracellular C-terminus (Jembrek et al. 2015). The central channel is formed by the M2 segments of each subunit. M1, M3 and M4 are designed to separate M2 from membrane lipids. Between M2 and M3 of the α-subunit at positions 270 and 291 there is a pocket which is responsible for binding alcohols and anesthetics. GABA molecules bind between the alpha and the beta subunit. It has been shown that beta subunits are mainly responsible for the ion selectivity of the receptor (Jembrek et al. 2015).

Alternative splicing provides for the enormous variety of GABAA receptor isoforms present. Currently, it is assumed that the main isoforms are composed of α1, β2 and γ2 subunits. The subunit composition characterizes the pharmacological properties of a given receptor. Due to their location, there are two types of GABA (A) receptors: synaptic and extrasynaptic. Their mutual relationship dynamically changes under the influence of various substances such as ethanol, GABA or neurosteroids. The sigma subunit, so far, is the only purely extrasynaptic GABAA receptor subunit (Jembrek et al. 2015).

GABAB receptors are heterodimers built of GABAB1 and GABAB2 subunits [4]. Each of them is composed of an N-terminal extracellular domain (VFT), a transmembrane domain (7TM) and an intracellular C-terminus. The

VFT module is constructed of an N-terminal lobe (LB1) and a C-terminal lobe (LB2). In inactive GABA receptors, the VFT domains of two different subunits are linked by three tyrosine residues at the LB1-LB1 interface. When the GABA receptor is activated, LB2 is also involved in the formation of intermolecular bonds. It has been suggested that this promotes intracellular signaling and increases the affinity of the agonist for the receptor. [4] Binding of an agonist to GABA1 alters the VFT pattern so that the linker regions are greatly approximated. VFT and 7TM are linked to each other by a linker region that is made up of about 20 amino acid residues. A characteristic feature is the lack of cysteine residues, which are found among other G protein-linked receptor linkers. This structure connects to ECL2 7TM by means of an anti-parallel B-sheet, which additionally stabilizes the entire structure (Jembrek et al. 2015).

Within each TM there are three extracellular loops (ECL) and three intracellular loops (ICL) that stabilize the connection between the VFT and the TM of two different subunits. [3] In an inactive state within the 7TM domain, there are interactions between TM3 (GABAB1) -TM5 (GABAB2) and TM5 (GABAB1) -TM3 (GABAB2). The TM3 / TM5 interface has been shown to be responsible for inhibiting receptor activity in the absence of an agonist. Receptor activation induces domain rotation, which creates a new TM6 interface from both monomers.

A phospholipid molecule is attached to the extracellular half of the 7TM region in both active and inactive form. Current research suggests that these molecules act as negative allosteric modulators (NAMs) to stabilize the inactive conformation. Phospholipids are believed to affect the ability to move by ECL2 into native GABA, modulating the activity of the receptor. In addition, binding to phospholipids is involved in the transmission of structural transitions from VFT to 7TM. Differences in the three-dimensional structure of GABAB1 and GABAB2 subunits determine the different ability to bind to molecules. GABAB1 binds to orthoseric ligands, while GABAB2 has a greater affinity for the G protein. Due to alternative splicing methods, there are many different variants within the GABAB1 subunit. The most abundant isoforms are GABA1a and GABA1b. Structurally, they differ in the presence of the so-called sushi 1 and sushi 2 domains on the GABA1a subunit in the N-terminus. In GABA receptors that are heterodimers, GABAB1 and GABAB2 are linked to each other by strong non-covalent bonds between the VFT of both subunits. On the other hand, in the case of GABA receptors belonging to oligomers, individual heterodimers are connected by definitely weaker and transient bonds (Naffaa et al. 2022).

GABAc (or GABAA -ρ) receptors are homomeric ion channels. However, some studies have shown differences in the structure of the ρ subunits, which resulted in their pseudoheteromeric structure. There are 3 types of ρ subunits: ρ1, ρ2, ρ3.

In the GABAA receptor molecule, three main sites for the attachment of molecules have been found. These include a GABA site for agonist or antagonist molecules, a picotoxin site on 7TM, and a benzodiazepine site on the ECD. The GABA binding site is at the interface between the alpha and beta subunits. The influence of the receptor subunit structure on the degree of ligand binding has been demonstrated. In particular, the δ subunit enhances the affinity for GABA agonists, while the γ subunit gives it a lower affinity. BZ sites are allosterically associated with both GABA sites and picotoxin sites, therefore the development of drugs that selectively target BZ sites is very limited. This area is located on the border of the alpha and gamma subunits and shows great heterogeneity. On this basis, sites 1, 2 and 3 were distinguished. Attaching the ligand to the BZ site may take place when its concentration is high or there is no GABA site. GABAA and PAM agonists inhibit the activation of neurons, promoting sedation. We distinguish here min. barbiturates and benzodiazepines, zolpidem, zaleplon and zopiclone, valerenic acid, picrotoxin. Selective PAMs include, among others. intravenous anesthetics, which are used to induce general anesthesia during surgery, and in lower concentrations can be used as sleeping pills and sedatives. The most commonly used anesthetics are etomidate and propofol (Nilsson et al. 2011). Also, compounds such as some progesterone metabolites are the PAM of the GABAA receptor. Therefore, they may have a sedative and hypnotic effect. Ganaxolone, which is a derivative of progesterone, is currently used to treat insomnia, anxiety and epilepsy (Tonon et al. 2020).

GABA receptor agonists activate the receptor and stabilize its closed conformation. Antagonists that enhance the open conformation of the VFT have the opposite effect. Antagonists only attach to the LB1 domain via hydrogen bonding. CGP54626 and SCH50911 are the two exceptions that bind to LB2 and show a much greater affinity for GABA (B) receptors than the other antagonists. Gamma-Aminobutyric Acid is currently the best known agonist of the GABA receptor. It binds to the VFT through both the LB1 and LB2 domains. GABA (B) receptor activity is regulated by positive allosteric regulators (PAM), negative allosteric regulators (NAM) and silent allosteric regulators (SAM). PAM includes such molecules as GS39783, rac-BHFF and BHF177. These molecules do not activate the receptor on their own but are involved in enhancing the response in the presence of an agonist. CLH304a is the most widely known NAM. It has the opposite effect of PAM, resulting in an inhibition of the receptor response. The PAM of the GABAB receptor includes min. CGP7930 or GS39783. These compounds show a strong anxiolytic effect. However, SAM has not yet been detected (Tonon et al. 2020).

GABAB receptor agonists can also cause sedation and sedation, and baclofen and its derivatives are a typical representative of this group of compounds. Baclofen is used as a muscle relaxant in the treatment of spasticity caused by trauma to the central nervous system. Its best-known derivative is gamma-hydroxybutyric acid (GHB), which also has the ability to activate GABA (B). It can be used to induce sedation and to reduce the symptoms of cataplexy in patients with naroplexy. Methylphosphinic acid agonists such as CGP44533 and CGP44532 are definitely more potent and longer acting than baclofen. The 3-aminopropylphosphonic acid derivatives CGP27492 and CGP35024 work in a similar way. GABAB receptor antagonists are primarily baclofen analogs - phaclofen, sacrofen and 2-hydroxysaclofen. Over time, other antagonists have also been developed that have been shown to penetrate the blood-cerebrospinal fluid barrier. These include: 3-aminopropyl-butylphosphinic acid, 3-aminopropyl-dietho-xymethylphosphinic acid, and 3-aminopropyl-cyclohexylmethyl-phosphinic acid.

GABA receptors play a role in the development of the body. It has been proven that in the second half of pregnancy and the first few months of a child's life, there are intense changes in the structure of the GABAergic system. This is essential for the proper development and function of the network of connections in the cortex of the brain and neural progenitor cells. There are also morphological changes in the structure of neurons. Lack of this early stimulation causes a significant reduction in the number of dendritic connections and the length of neurites. The strong protective effect of the GABA system against oligodendrocytes has also been demonstrated. The expression of GABA receptors is strongly marked in the areas of the brain responsible for learning and memory. Due to their presence in the mesolimbic system, the symptoms of addiction are alleviated by inhibiting the dopaminergic pathways. GABA also exhibit a number of activities outside the central nervous system. They induce the conversion of $\alpha$ to $\beta$ cells in the pancreas, as well as their proliferation. These receptors are also found on B and T lymphocytes, dendritic cells and macrophages, and their stimulation reduces the proliferation of T lymphocytes and changes the profile of their cytokines (Tonon et al. 2020).

In studies carried out in mice, GABAB receptors have been shown to play an important role in emotional disorders such as anxiety and depression. The R1 subunit deficient mice were definitely more restless than their counterparts. GABAergic deficits were also discovered in the course of schizophrenia. It has been proven that in Alzheimer's disease there are changes in the expression and activity of the GABA receptor, but also in the remodeling of its entire system with focal accumulation of high concentrations of GABA in the postmortem tissue. The scheme of the functioning of the GABA system in neurodevelopmental disorders is very complicated and its influence is still not fully understood. Current research suggests that autism, Angelman, Rett syndrome, and fragile X chromosome have elements of GABAergic insufficiency. On the other hand, it is hyperactive in Down's syndrome. Changes in the R1 subunit also increase the risk of temporal epilepsy, and its deficiency leads to generalized seizures. Stimulation of GABA receptors by benzodiazepines or anticonvulsants is also used to alleviate alcohol withdrawal symptoms and alcoholism itself. Rapid changes in the GABA system were also observed during stress reactions. There was an increase in the number of GABA receptors and their increased activity in many cancer cell lines originating, inter alia, from the liver, colon, stomach and thyroid cells. Changes in neuronal excitability that are induced by activation of the GABA system are essential for the functional regeneration of the cerebral cortex in the post-stroke period. Conditions such as stroke increase the transmission and activation of GABA receptors in the infarcted area.

GABA receptors are a heterogeneous group of receptors within which, thanks to alternative splicing methods, there is considerable diversity in subunit structure. On the one hand, this dissimilarity determines the differentiated localization both within the central nervous system and other organs, and on the other hand, it is a condition for clinically significant ligand binding specificity. Due to their wide distribution in the central nervous system, they play a number of important functions in it. The influence of GABA receptors on the functioning of the body begins in fetal life, where they ensure the correct process of shaping neural networks. Later, they take part in such processes as learning, memory, and shaping emotions and appropriate behavior. They perform the latter function thanks to the high concentration within the structures of the mesolimbic system. It is necessary to know the location of individual types of receptors and their functioning, since many of them play a significant role in the pathomechanism of diseases of the central nervous system. Currently, numerous studies are carried out on how to modify them and its possible application in medical practice.

## References

Ghit A, Assal D, Al-Shami AS, Hussein DEE. $GABA_A$ receptors: structure, function, pharmacology, and related disorders. *J Genet Eng Biotechnol.* 2021; 19(1): 123.

Jembrek MJ, Vlainic J. GABA Receptors: Pharmacological Potential and Pitfalls. *Curr Pharm Des.* 2015; 21(34): 4943-59.

Naffaa MM, Hibbs DE, Chebib M, Hanrahan JR. Pharmacological Effect of GABA Analogues on GABA-$\varrho$2 Receptors and Their Subtype Selectivity. *Life (Basel).* 2022 Jan 17;12(1):127.

Nilsson J, Sterner O. Modulation of GABA(A) receptors by natural products and the development of novel synthetic ligands for the benzodiazepine binding site. *Curr Drug Targets.* 2011; 12(11): 1674-88.

Tonon MC, Vaudry H, Chuquet J, Guillebaud F, Fan J, Masmoudi-Kouki O, Vaudry D, Lanfray D, Morin F, Prevot V, Papadopoulos V, Troadec JD, Leprince J. Endozepines and their receptors: Structure, functions and pathophysiological significance. *Pharmacol Ther.* 2020; 208: 107386.

# Chapter 22

# PTX Receptors

**Jadwiga Inglot**
**Dorota Bartusik-Aebisher***
**and David Aebisher**
Medical College of the University of Rzeszów, Poland

## Abstract

Bacterial ADP-ribosylation toxins constitute a large family of dangerous toxins, including pertussis, cholera and diphtheria toxins, which, as cytotoxic agents, cause severe infectious diseases, including whooping cough, cholera, and diphtheria. Among these toxins, Pectenotoxin (PTX) is the predominant. It catalyses ADP ribosylation of the α-subunits of the Gi/o family of heterotrimeric proteins (Gαi, Gαo and Gαt), thus preventing interaction of the G proteins with their cognate G protein coupled receptors (GPCR). PTX plays a key role in the pathogenesis of whooping cough, the development of protective immunity against reinfection, and is an essential component of new acellular vaccines.

**Keywords:** PTX, PTX receptors, G-protein-coupled receptor

Pectenotoxin is a class A-B exotoxin. It consists of 952 residues forming six subunits (S1, S2, S3, two copies of S4 and S5). It contains two functionally distinct domains that differ in both their sequence quaternary structure and pathogenic mechanisms. The A domain (S1 subunit) is responsible for the catalytic activity of the toxin (ADP-ribosyltransferase activity), leading to the

* Corresponding Author's Email: dbartusikaebisher@ur.edu.pl.

In: The Medical Biology Guide to Proteins
Editor: David Aebisher
ISBN: 979-8-88697-910-7

inhibition of receptor-G protein coupling. The B domain consists of subunits from S2 to S5 and binds carbohydrate-containing receptors that allow the toxin to enter cells (deliver the A domain to the cytosol). The ADP-domain A-ribosylates the α subunits of heterotrimeric Gi/o proteins (Gαi/o), whereby the receptors are detached from the Gi/o proteins. This modification of the Gαi/o proteins results in increased accumulation of cAMP, which is one of the mechanisms by which PTX induces various pathological effects in host cells (Costa et al. 2018).

The B domain binds cell surface expressed proteins, such as the Toll-like 4 receptor, and activates the intracellular signal transduction cascade. The S1 subunit (domain A) has been shown to have about 20% sequence identity to the A subunit of cholera toxin. The amino acid sequences of S2 and S3 are 70% identical, but no other sequence homology was found between the various B domain subunits of PTX or between these sequences and the B subunits of other bacterial toxins. While many of the effects of PTX are mediated by ADP ribosylation of Gαi/o proteins, a GI/o protein independent effect of PTX has also been reported. Thus, PTX modifies cellular responses through at least two different signaling pathways; ADP-ribosylation of Gαi/o proteins by A domain (Gi/o protein dependent action) and B domain interaction with cell surface proteins (Gi/o protein independent action) (Das et al. 2005). PTX plays an important role in the development of protective immunity against whooping cough and is an essential component of new acellular vaccines. It is also widely used as a biochemical tool for the ADP-ribosylate of GTP binding proteins in signal transduction studies (Mangmool et al. 2011).

Upon attachment of PTX to host cells, the S2 and S3 subunits of the B domains bind to exposed sugar glycolipid (gangliosides) residues on the host cell membrane. The A domain (S1 subunit) crosses the membrane and is released from the B domain into the cytoplasm. Upon entry into the cell, the A domain ribosylates specific target proteins such as the α subunit of heterotrimeric Gi/o proteins through its ADP-ribosyltransferase activity. PTX catalyzes the cleavage of the C-N bond between the ribose carbon and the nicotinamide nitrogen and transfers the ADP-ribosyl moiety from nicotinamide adenine dinucleotide ($NAD^+$) to an acceptor molecule on the target protein. ADP-ribosylation of the Gαi/o proteins prevents conjugation to their cognate GPCRs and consequently disrupts the signaling cascade. The ADP-ribosylated amino acid by PTX is cysteine, which is four residues from the carboxy terminus of the α-subunits of Gi/o proteins. Separation of GPCR from Gαi/o proteins disrupts communication between the receptor and the AC effector molecule. Thus, the Gαi/o protein is inactivated and cannot perform its normal AC inhibitory function. Thus, it prevents the signal from the Gi/oPCR. The conversion of ATP to cAMP cannot be stopped, resulting in excessive intracellular cAMP levels and the subsequent disruption of many cellular processes (Sim et al. 2017).

Whooping cough is an important cause of infant morbidity and mortality worldwide, accounting for approximately 400,000 deaths each year. Disease control with vaccines has been threatened by uncertainties about the safety of traditional inactivated B-cell vaccines.

The determination of the PTX crystal structure helped to obtain detailed information on the functional and immunological determinants needed to refine the design of recombinant PTX molecules suitable for use as components of acellular vaccines. The role of PTX in common pertussis disease in other people is less clear, but substantial evidence of its involvement in pathogenesis has been gathered from animal model studies. Understanding the structure of PTX is also a step towards understanding how PTX interacts with eukaryotic cells and modifies their functions. Comparison of the structure of PTX with the known structures of other bacterial A-B toxins suggests how they may have evolved. The ADP-ribosylation of Gai/o proteins does not explain all possible outcomes of PTX effects with host cells. Further research on the structure, operation and dysfunction of both PTX and its receptors will help to better understand the pathomechanism of whooping cough development and will allow for more effective protection against its development.

G protein-coupled receptors convert the signal from extracellular ligands to intracellular responses mediated by heterotrimeric G proteins. Heterotrimeric G proteins are so defined by their three subunits: α, β and γ, the former being a GTP-binding protein with GTPase activity, and the latter two form a single, inseparable complex, usually referred to as the Gβγ subunit. G proteins are divided into four families according to the functional and structural homologues of their Gα subunits, which are responsible for the major properties of the Gαβγ tri-complex: Gi/o, Gs, Gq/11 and G12/13 (Ågren et al. 2020). Pertussis toxin (PTX) is an ADP-ribosylation toxin. PTX catalyzes ADP ribosylation of the α-subunits of the Gi/o family of heterotrimeric proteins (Gαi, Gαo and Gαt; except for Gαz), thus preventing the interaction of G proteins with their cognate G protein coupled receptors (GPCR).

The Gi/o family is encoded by 8 genes. The Gi/o proteins are generally described as inhibitory, with adenylyl cyclase (AC) and potassium channels as major effectors. The three Gi proteins: Gαi1, Gαi2 and Gαi3 inhibit some isoforms of AC ("i" refers to its inhibitory effect), reducing the ability of basal and Gs stimulated adenylyl cyclase to produce cAMP. Gt1/2 (transducin) and Ggust (gustducin) are involved in visual and taste functions, respectively, and activate cGMP-phosphodiesterase (cGMP-PDE). Gz (which is not affected by PTX) inhibits AC, stimulates potassium channels, and interacts with several RGS (regulator of G protein signaling) proteins. Gz is also phosphorylated by protein kinase C (PKC) and p21 1 activated kinase (PAK1). The Go protein ("o" stands for "other") was discovered while cleaning Gi from the brains of cattle. The Go gene transcript (GNAO) is alternatively assembled into Gα1 and Gα2 variants, and some studies suggest that the ability of Go to modulate AC activity is mainly due to the less studied Go2 isoform or its βγ subunit. Go also appears to play a significant role in modulating several other signaling pathways, including the STAT3 and ERK pathways (Ågren et al. 2020).

All proteins belonging to the Gi/o family, with the exception of the Gz protein, are inactivated by the pertussis toxin. Gi/o-coupled GPCRs stimulate GIRK channels through direct interactions between Gβγ and the channel, leading to hyperpolarization and thus reduction of neuronal excitability. Importantly, while Go is the most abundant heterotrimeric G protein in the central nervous system (CNS), it makes up about 1% of the total membrane protein in the brain. Very little is known about its function, including what is particularly true of the Gα2 isoform (Costa et al. 2018).

The function, expression and number of amino acids of each protein are given in the table.

**Table 1.** Gαi/o protein family

| The α subunit | The number of amino acids | Expression | Effector effect |
|---|---|---|---|
| αi1-αi3 | 354 | Neurons and ubiquitous | Inhibition of AC activity<br>Inhibition of calcium channels<br>Activation of potassium channels |
| αt | 350 | The outer parts of the suppository and the rod | Activation of cGMP-PDE |
| αgust | 353 | Taste buds: sweet/ bitter | Activation of cGMP-PDE |
| αz | 355 | Platelets | Inhibition of AC activity<br>Inhibition of calcium channels<br>Activation of potassium channels |
| αo | 354 | Heart, neurons, neuroendocrine cells | Inhibition of AC activity<br>Inhibition of calcium channels<br>Activation of potassium channels |

## References

Ågren, R., and Sahlholm, K. (2020). Voltage-Dependent Dopamine Potency at $D_1$-Like Dopamine Receptors. *Front Pharmacol.*, 11, 581151.

Costa, R., Bicca, M. A., Manjavachi, M. N., Segat, G. C., Dias, F. C., Fernandes, E. S., and Calixto, J. B. (2018). Kinin Receptors Sensitize TRPV4 Channel and Induce Mechanical Hyperalgesia: Relevance to Paclitaxel-Induced Peripheral Neuropathy in Mice. *Mol Neurobiol.*, 55(3), 2150-2161.

Das, P., and Dillon, G. H. (2005). Molecular determinants of picrotoxin inhibition of 5-hydroxytryptamine type 3 receptors. *J Pharmacol Exp Ther.*, 314(1), 320-8.

Mangmool, S., and Kurose, H. (2011). G(i/o) protein-dependent and -independent actions of Pertussis Toxin (PTX). *Toxins (Basel).*, 3(7), 884-99.

Sim, Y. B., Park, S. H., Kim, S. S., Lim, S. M., Jung, J. S., Sharma, N., and Suh, H. W. (2017). Spinal β-adrenergic receptors' activation increases the blood glucose level in mice. *Anim Cells Syst (Seoul).*, 21(4), 278-285.

# Chapter 23

# Estrogen Receptors

**Julia Kudła**
**Dorota Bartusik-Aebisher***
**and David Aebisher**
Medical College of the University of Rzeszów, Poland

## Abstract

The estrogen receptors have many important functions during the development and maturation of tissues. Their expression is not only related to the reproductive system, but also occurs in the lungs, prostate, cardiovascular and nervous systems. The results of many studies carried out on their specification and mechanism of action have made it possible to create many therapies for various diseases in which estrogen receptors are involved in their formation.

**Keywords:** estrogens, receptors, DNA, tamoxifen, clomiphene, toreomiphene, estran

Estrogens group includes estrone, estradiol, estriol and estetrol. They are commonly called female hormones, which, however, does not contradict their occurrence in men, in which they also play an important role, e.g., in liquefying sperm. These hormones interact by binding to specific estrogen receptors that stimulate transcriptional processes or signaling events, resulting in the control of gene expression. This can be done by direct binding of

---

* Corresponding Author's Email: dbartusikaebisher@ur.edu.pl.

In: The Medical Biology Guide to Proteins
Editor: David Aebisher
ISBN: 979-8-88697-910-7

estrogen receptor complexes to specific sequences in gene promoters, we say then having a genomic effect or processes that do not involve direct binding to DNA (non-genomic effects). As we can see, the influence of estrogens on gene expression is regulated by complex mechanisms, regardless of whether these hormones affect cells through nuclear or non-nuclear effects (Chantalat et al. 2020).

Estrogen receptors were discovered as the first receptors to bind to a given hormone in 1958 by Elwood Jensen. He showed that estrogens are taken up from the blood by proteins in female sex cells. Ten years later, he discovered that receptors, when bound to estrogen, migrate to the nucleus where they affect gene expression. It has since been argued that there is only one type of estrogen binding receptor ERα. It has changed in 1996, when a team of researchers led by Dr. Jan-Ake Gustafsson demonstrated the existence of Erβ proteins in prostate secretory epithelial cells and in ovarian granulosa cells that were highly homologous to previously discovered Erα. In 2012, the existence of estrogen receptors coupled with GPER1 G proteins was detected by molecular cloning methods (Chen et al. 2008).

Among estrogen receptors, we distinguish ER-alpha and ER-beta nuclear receptors and GPER1 membrane receptors. Nuclear receptors are encoded by two different genes, but their structure is identical, however, some structural elements have different degrees of similarity. The ER-alpha receptor is encoded by the ESR1 gene located in the q24-q27 part of the sixth autosome. The full-size Er-alpha isoform has a mass of 66 kDa and consists of 595 amino acids, in addition to it, a few shorter ones with a mass of 36 kDa and 46kDa have been detected. The second type, i.e., Er-beta, is encoded by the ESR2 gene in the 14q23-24 of the fourteenth chromosome. It is made of 530 amino acids and its mass is 59 kDa. ER-beta and alpha also contain several isoforms (54 kDa, 49 kDa, and 44 kDa. As mentioned earlier, alpha and beta estrogen receptors consist of five domains with specific functions and different homology. against ERalpha ERbeta. The A/B domain is responsible for the ligand-independent but tissue-specific and promoter-specific transcriptional activity of the receptor; it binds to the transcription complex or contacts the coregulatory, DBD (DNA binding domain) - its function is specific DNA binding and receptor dimerization the hinge domain is responsible for the proper localization in the nucleus, it also enables conformational changes of the molecule. LBD (ligand binding domain) made of 12 α-helices, contains a ligand binding site which results in transcription. It binds to the HSP protein and is responsible for dimerization C-terminal domain activates transcription (Eyster et al. 2016).

Estrogen is a steroid hormone; therefore it crosses the membrane and binds to intracellular ER-alpha and ER-beta, and then interacts by binding to DNA sequences. This hormone can also trigger a cascade of reactions as a result of its interaction with GPER1 or ER-beta and alpha. As a result of differences between cellular responses that lead to the regulation of gene expression as a result of the interaction of the receptor-estrogen complex indirectly or directly from DNA, these events were divided into genomic (classical) and non-genomic. In the genomic mechanism, estrogen binds to the receptor, then this complex is translocated to the nucleus (Heldring et al. 2007).

In the next stage, the receptor dimerizes (alpha-alpha and beta-beta homodimers and alpha-beta heterodimers) and connects to a specific DNA sequence, i.e., element of the ERE response (estrogen response element), which is located in the promoter of specific genes. At the same time a combination of estrogen with the receptor, it activates conformational changes in the ligand-binding receptor, which enables the binding of coactivators. Estrogen hormones can also function without direct initiation of target gene transcription and protein synthesis, this is called a non-genomic mechanism which is faster than the classical response model. In a non-genomic model, estrogen binds to the membrane receptor GPER1 and can then activate the synthesis of cAMP, protein kinases in signaling cascades that cause indirect changes in gene expression, influencing cellular metabolism. The activation of these receptors depends on the concentration of free hormone or other ligands in the blood, their affinity for the receptor, as well as changes induced after previous activation. In addition to estrogens, which are produced by the gonads, there are also organic and inorganic compounds that can recognize and bind to estrogen receptors. ER-alpha in relation to ER-beta is characterized by higher selectivity. There are major classes of estrogen receptor ligands. These are: endoestrogens, xenoestrogens, phytoestrogens, metalloestrogen, SERM. The latter group of ligands is special because they both act as agonists and antagonists depending on the tissues. The SERMs include tamoxifen, clomiphene, and toreomiphene. Tamoxifen is one of the most commonly used drugs in the treatment of breast cancer because it acts as an antagonist on the receptors of the mammary gland cells and has an agonist effect on uterine cells (Heldring et al. 2007).

As a result of many clinical trials, it has been proved that one of the risk factors for the development of breast cancer is the increased exposure of the mammary gland epithelium to estrogen. Estrogen comes together with estrogen receptors, stimulates the proliferation of breast cells, intensification of cell division, DNA synthesis increases the risk of errors during replication, which may result in the acquisition of harmful mutations that interfere with normal cellular processes, e.g., apoptosis, cell proliferation or DNA repair. It has also been shown that after the end of treatment, remission of breast cancer occurred in about 60% of patients whose cancer cells had ER expression, more often when the progestrogen receptor PgR was present (Nilsson et al. 2001).

As we know, estrogen prevents osteoporosis in both sexes because it inhibits the activity of osteoclasts and increases osteoblasts, causing the formation of bone tissue. This hormone is also believed to prevent postmenopausal bone loss. Despite the many results confirming the association of ERalpha and ERbeta polymorphisms with the occurrence of osteoporosis, there is some controversy over this. However, it has been shown that patients with mutations in the gene encoding ERalpha exhibit incomplete epiphyseal closure and decreased bone mineral density (Nilsson et al. 2001).

Estrogen receptors, which are the subject of this chapter, are involved in the regulation of numerous physiological processes. They influence cellular metabolism through genomic or non-genomic mechanisms as a result of cascades of intracellular reactions. By binding with specific ligands, they affect many different tissues in the body and mutations in the genes encoding them can cause many dysfunctions. The results of many studies carried out on their specification and mechanism of action have made it possible to create many therapies for various diseases in which estrogen receptors are involved in their formation.

## References

Chantalat E, Valera MC, Vaysse C, Noirrit E, Rusidze M, Weyl A, Vergriete K, Buscail E, Lluel P, Fontaine C, Arnal JF, Lenfant F. Estrogen Receptors and Endometriosis. *Int J Mol Sci*. 2020; 21(8): 2815.

Chen GG, Zeng Q, Tse GM. Estrogen and its receptors in cancer. *Med Res Rev*. 2008; 28(6): 954-74.

Eyster KM. The Estrogen Receptors: An Overview from Different Perspectives. *Methods Mol Biol*. 2016; 1366: 1-10.

Heldring N, Pike A, Andersson S, Matthews J, Cheng G, Hartman J, Tujague M, Ström A, Treuter E, Warner M, Gustafsson JA. Estrogen receptors: how do they signal and what are their targets. *Physiol Rev*. 2007; 87(3): 905-31.

Nilsson S, Mäkelä S, Treuter E, Tujague M, Thomsen J, Andersson G, Enmark E, Pettersson K, Warner M, Gustafsson JA. Mechanisms of estrogen action. *Physiol Rev*. 2001; 81(4): 1535-65.

# Chapter 24

# Ricin

**Federica Adamo**
**Dorota Bartusik-Aebisher***
**and David Aebisher**
Medical College of the University of Rzeszów, Poland

## Abstract

Ricin is a potent toxin derived from the castor plant, Ricinus communis L. Ricin intoxication mimics a variety of disease states, thus a low threshold of suspicion must be maintained to recognize a potential epidemic. The castor plant, native to the southeastern Mediterranean region, eastern Africa, and India, it is now widespread throughout temperate and subtropical regions. Experimental animal studies reveal that clinical signs and pathological manifestations of ricin toxicity depend on the dose as well as the route of exposure. Contact with ricin powders or products may cause redness and pain of the skin and the eyes. Underway to develop small molecule inhibitors for the treatment of ricin intoxication. Recent findings suggest that refinement of the newly identified ricin inhibitors will yield improved compounds suitable for continued evaluation in clinical trials.

**Keywords:** ricin, ricin toxin A chain (RTA), antiinflammatory agents, oral intoxication

---

* Corresponding Author's Email: dbartusikaebisher@ur.edu.pl.

In: The Medical Biology Guide to Proteins
Editor: David Aebisher
ISBN: 979-8-88697-910-7

Ricin is a potent toxin derived from the castor plant, Ricinus communis L (Euphorbiaceae). The purified ricin toxin is a white powder that is water soluble; it inhibits protein synthesis leading to cell death. Ricin intoxication mimics a variety of disease states, thus a low threshold of suspicion must be maintained to recognize a potential epidemic and the treatment is largely symptomatic and supportive. In 1978, the lethality of ricin was overtly established after the high-profile assassination of Bulgarian dissident Georgi Markov, later other similar events followed, regarding ricin's potential for urban bioterrorism, and thus prompted its constant inclusion in weapons of mass destruction investigations (Abbes et al., 2021). The quaternary structure of ricin is a globular, glycosylated heterodimer that consists of a 32 kilodalton A chain glycoprotein linked by a disulfide bond to a 32 kilodalton B chain glycoprotein.

The Ricin toxin A chain (RTA) is an alpha/beta protein which contains eight alpha helices and eight beta sheets. It has three domains. Domain 1 consists of a beta sheet containing both parallel and anti-parallel strands. The second alpha helical domain makes up the core of the protein, and includes the active site responsible for inactivating the Ribosome via depurination. The third domain contains a helix plus two beta strands and interacts with the Ricin toxin B chain (RTB), a lectin that binds to galactose-containing surface receptors (Bozza et al., 2015) .

The castor plant, native to the southeastern Mediterranean region, eastern Africa, and India, it is now widespread throughout temperate and subtropical regions. Object of study by Peter Hermann Stillmark, a student at the Dorpat University in Estonia, who discovered ricin in 1888 and observed how it caused agglutination of erythrocytes and precipitation of serum proteins. In 1891, Paul Ehrlich studied ricin and abrin in pioneering research that is recognized as the foundation of immunology, he found that animals vaccinated with small oral doses of castor beans were protected against a lethal dose of the toxin. Ricin, contained in Castor oil, was reportedly used as an instrument of coercion by the Italian Squadristi, the Fascist armed squads of Benito Mussolini against political dissidents and regime opponents that were forced to ingest large amounts of castor oil, triggering severe diarrhea and dehydration that often led to death (Franke et al., 2019).

Experimental animal studies reveal that clinical signs and pathological manifestations of ricin toxicity depend on the dose as well as the route of exposure. The common routes of entry are oral intoxication (ingestion), injection, and inhalation. The differences observed in pathology among various routes likely result from the fact that RTB binds to a wide array of cell surface carbohydrates. Once bound, RTA is internalized and results in the death of intoxicated cells. Additionally, in animals and humans intoxicated either by injection or oral ingestion, a transient leukocytosis is commonly observed, with leukocyte counts rising two to five times above their normal values. Within a few hours of inhaling significant amounts of ricin, the likely symptoms would be respiratory distress, fever, cough, nausea, and tightness in the chest. Heavy sweating may follow as well as fluid building up in the lungs (pulmonary edema). This would make breathing even more difficult, and the skin might turn blue. Finally, low blood pressure and respiratory failure may occur, leading to death. If someone swallows a significant amount of ricin, he or she would likely develop vomiting and diarrhea that may become bloody. Severe dehydration may be the result, followed by low blood pressure. Other signs or symptoms may include seizures, and blood in the urine. Within several days, the person's liver, spleen, and kidneys might stop working, and the person could die. Ricin is unlikely to be absorbed through normal skin. Contact with ricin powders or products may cause redness and pain of the skin and the eyes. However, if you touch ricin that is on your skin and then eat food with your hands or put your hands in your mouth, you may ingest some. Death from ricin poisoning could take place within 36 to 72 hours of exposure, resulting from a severe inflammatory response and multiorgan failure (Franke et al., 2019).

Despite the history of ricin's use as a weapon, and unlike other toxin-mediated illnesses, no Food and Drug Administration-approved or vaccine against ricin intoxication exists. Given that ricin does not have cell specific selectivity, treatment of ricin intoxication is dependent on the site or route of entry, is largely symptomatic, and basically supportive to minimize the poisoning effects of the toxin. Medical countermeasures that have demonstrated capability to disrupt the ricin intoxication process include vaccines and antibody therapy. Both rely on the ability of antibodies to prevent the binding of ricin to cell receptors. To ensure maximum protection, the vaccine must be given before exposure, and sufficient antibody must be produced. The route of exposure for any agent is an important consideration in determining prophylaxis and therapy. For oral intoxication, supportive therapy includes intravenous fluid and electrolyte replacement and monitoring of liver and renal functions. Standard intoxication principles should be followed. Because of the necrotizing action of ricin, gastric lavage or induced emesis should be used cautiously. An aerosol-exposed patient may require the use of positive-pressure ventilator therapy, fluid and electrolyte replacement, anti-inflammatory agents, and analgesics. Percutaneous exposures necessitate judicious use of intravenous fluids and monitoring for symptoms associated with VLS (Lord et al., 2003).

Since vaccination offers a practical prophylactic strategy against ricin exposure, considerable efforts have been devoted to developing a safe and effective ricin vaccine to protect humans, in particular soldiers and first responders. Recombinant candidate ricin vaccines are currently in advanced development in clinical trials. Efforts are also underway to develop small molecule inhibitors for the treatment of ricin intoxication. Recent findings suggest that refinement of the newly identified ricin inhibitors will yield improved compounds suitable for continued evaluation in clinical trials.

The United States looked at using ricin as a weapon in World War I, either using it to coat bullets, or creating a dust cloud of the toxin that would be inhaled by the enemy. However, ricin's sensitivity to heat made its use on bullets problematic, and the dust cloud idea was dismissed until an antitoxin could be found. In World War II, British, French, Canadian, and American scientists all studied the possibility of using ricin as a weapon, and the U.S. conducted tests at the Dugway Proving Ground in Utah in 1944; however, it was not used in World War II, and it was set aside while more deadly toxins such as botulinum toxin A were favored as a biological weapon. Because of its limitations, ricin seemed more appropriate for small-scale or personal attacks, becoming a weapon for terrorists (Spooner et al., 2015).

On September 7, 1978, in London, outspoken Bulgarian dissident Georgi Markov was stabbed in the leg with an umbrella by an unknown foreign man. Within three days he was dead. It was found that the tip of the umbrella had held a miniscule metal sphere containing a pellet of ricin that remained in the wound and killed him. Both the Bulgarian secret services and the KGB were suspected of being behind this assassination. The affair has remained famous within the London police and is known as "The Bulgarian Umbrella" and since then, there have been several attempted uses of ricin as a terrorist weapon. Although ricin's potential use as a military weapon was investigated, its utility over conventional weaponry remains ambiguous. Despite its toxicity, ricin is less potent than other agents such as botulinum neurotoxin or anthrax. Furthermore, wide-scale dispersal of ricin is logistically impractical. Thus, while ricin is relatively easy to produce, it is not as likely to cause as many casualties as other agents.

## References

Abbes M, Montana M, Curti C, Vanelle P. Ricin poisoning: A review on contamination source, diagnosis, treatment, prevention and reporting of ricin poisoning. *Toxicon.* 2021; 195: 86-92.

Bozza W P, Tolleson W H, Rivera Rosado LA, Zhang B. Ricin detection: tracking active toxin. *Biotechnol. Adv.* 2015; 33(1): 117-123.

Franke H, Scholl R, Aigner A. Ricin and Ricinus communis in pharmacology and toxicology-from ancient use and "Papyrus Ebers" to modern perspectives and "poisonous plant of the year 2018." *Naunyn Schmiedebergs Arch. Pharmacol.* 2019; 392(10): 1181-1208.

Lord M J, Jolliffe N A, Marsden C J, Pateman C S, Smith D C, Spooner R A, Watson P D, Roberts L M. Ricin. Mechanisms of cytotoxicity. *Toxicol Rev.* 2003; 22(1) :53-64.

Spooner R A, Lord J M. Ricin trafficking in cells. *Toxins (Basel).* 2015; 7(1): 49-65.

# Chapter 25

# Nicotinic Acetylcholine Receptors

**Karol Bednarz**
**Dorota Bartusik-Aebisher***
**and David Aebisher**
Medical College of the University of Rzeszów, Poland

## Abstract

The nicotinic acetylcholine receptors belong to the group of acetylcholine responsive polypeptide receptors. They are ligand-gated ion channels and can be divided into two groups: muscle receptors, which are found at the neuromuscular junction of skeletal muscles where they mediate neuromuscular transmission, and neuronal receptors, which are found throughout the peripheral and central nervous systems. In the peripheral nervous system, they transmit signals from presynaptic cells to postsynaptic cells in the sympathetic and parasympathetic nervous systems. A single receptor is pentametric around a water-filled pore. The compounds with the ability to activate prescription acetylcholine include, among others nicotine, epibatidine, anabasein, α anotoxin, arecoline, lobeline or cytisine. Activation of nicotinic receptors leads to two main mechanisms. The first is the flow of cations through the receptor which depolarizes the cell membrane resulting in an excitatory postsynaptic potential in neurons, leading to the activation of voltage-gated ion channels. The second effect is the influence of $Ca^{2+}$ ions, which indirectly affects various intracellular cascades, which leads to the regulation of the activity of certain genes or the release of neurotransmitters.

**Keywords:** nicotinic receptor, nicotine, synapse, cytisine

---

* Corresponding Author's Email: dbartusikaebisher@ur.edu.pl.

In: The Medical Biology Guide to Proteins
Editor: David Aebisher
ISBN: 979-8-88697-910-7

The nicotinic acetylcholine receptors are ligand-gated homo- or heteropentameric ion channels of the Cys-loop superfamily. The mass of one receptor is 290 kDa. They are composed of α subunits (α1 to α10), β subunits (β1-β4) and γ, δ and ε subunits. Most receptors contain one type of α subunit and one type of β subunit. The most common combinations are the α4β2 heteromeric and α7 homomeric subtypes. These two receptors are the most dominant subtypes found in the brain and the most frequently targeted receptor subtypes in drug discovery programs to date. Much of the structural and functional diversity of nicotinic receptors is due to the many possible combinations of subunits. All subunits share a common architecture consisting of a large N-terminal extracellular domain followed by four hydrophobic transmembrane domains, a large cytoplasmic loop, and a short extracellular carboxyl domain. The transmembrane domains are arranged in concentric layers around a central pore filled with water. The large extracellular domain contains agonist binding sites. The transmembrane domain forms a water-filled channel that forms a hydrophilic ion pathway through the lipid bilayer membrane when the pores are open (Matta et al. 2021). The intracellular domain which is the most variable among the subunits and contains sites for modulation and interaction with cytoplasmic elements. The functional properties of each subtype are unique but overlap sufficiently to make them very difficult to distinguish by pharmacological agents, especially when the subtypes share subunits or contain different subunits with a high degree of homology. The nicotine receptor can be classified into the muscle type and the neuronal type. The muscle type is found in the membranes of neurons, while the muscle type is found in the neuromuscular synapses. Additionally, this receptor is found in vascular endothelial cells, airway and bronchial cells, respiratory fibroblasts, keratinocytes, oral epithelial cells, esophageal epithelial cells, astrocytes, urinary tract cells, lymphocytes, monocytes, macrophages, bone eosinophils, bone marrow cells, synoviocytes and placental cells (Changeux et al. 2020).

The opening of the ion channel occurs after the ligand binds to the receptor. In muscle-type nicotinic receptors, the ligand binding sites are located on the border of the α and ε or δ subunits. In neuronal nicotinic receptors, the binding site is at the interface with the α and β subunits or between two α subunits in the case of the α7 subunit. The binding site is in the extracellular domain near the N-terminus. When an agonist binds to a site, all subunits present undergo a conformational change and a channel is opened that forms a pore approximately 0.65 nm in diameter. Under normal physiological conditions, the receptor needs exactly two ligand molecules, acetylcholine, to open. Opening the channel allows positively charged ions to move through it. In addition to the endogenous acetylcholine agonist, nicotinic acetylcholine receptor agonists include nicotine, epibatidine, choline, anabasein, α anotoxin, arecoline, lobeline, and cytisine. All nicotinic receptor subtypes are permeable to monovalent ions such as $Na^+$ and $K^+$ and to $Ca^{2+}$ ions. By modulating the flow of cations such as $Na^+$, $K^+$ and $Ca^{2+}$ across cell membranes, nicotinic receptors regulate the excitability of neurons and the release of neurotransmitters, influencing many physiological processes, including behavior. Prolonged or repeated exposure to a stimulus often results in a diminished nicotinic receptor ligand response, known as desensitization. Desensitized receptors may revert to a prolonged open state when the agonist is bound in the presence of a positive allosteric modulator. Receptor function can be modulated by phosphorylation and activation of second messenger-dependent protein kinases. The nicotinic receptor can function according to 2 main mechanisms. Classic synaptic transmission relies on the release of high concentrations of neurotransmitters that act on directly adjacent receptors. Paracrine transmission involves neurotransmitters released by axon terminals which then diffuse through the extracellular environment until they reach receptors that may be distant (Hoskin et al. 2019).

A particular ligand for the nicotinic acetylcholine receptor is cytisine, which is used in the treatment of tobacco dependence. Cytisine is obtained from the seeds of Laburnum anagyroides and is very similar to nicotine. Cytisine competes with nicotine for the nicotinic receptor of the α4β2 subtype. The stimulation of these receptors opens a channel through which calcium or sodium ions enter the cell, which leads to the release of dopamine and other neurotransmitters, resulting in the intensification of the pleasure effect, reduction of anxiety and tension, improved memory, stimulation and decreased appetite. Cytisine has a 7-fold greater affinity for the α4β2 receptor and leads to the stimulation of dopamine secretion while being a nicotine antagonist. Cytisine, which is a strong nicotinic receptor agonist, stimulates the respiratory and vasomotor center, increases the secretion of adrenaline, increases blood pressure, reduces symptoms occurring in the period after nicotine withdrawal, making it a very good drug in the treatment of tobacco addiction (Martinez-Pena et al. 2021).

The nicotinic acetylcholine receptor is present throughout the body, the greatest number of receptors is found in the central and peripheral nervous system and in the neuromuscular junction. It consists of 5 α, β, γ, δ and ε subunits forming a water pore which is also an ion channel. A naturally occurring agonist in the body is acetylcholine, while nicotine is an exogenous agonist. In the treatment of nicotine addiction, cytisine is used as a partial agonist and antagonist at the nicotinic receptor. Further research on cytisine and an increasingly better understanding of the nicotinic acetylcholine receptor may significantly contribute to the treatment of nicotine addiction (Rueda Ruzafa et al. 2021).

## References

Changeux JP. Discovery of the First Neurotransmitter Receptor: The Acetylcholine Nicotinic Receptor. *Biomolecules*. 2020; 10(4): 547.

Hoskin JL, Al-Hasan Y, Sabbagh MN. Nicotinic Acetylcholine Receptor Agonists for the Treatment of Alzheimer's Dementia: An Update. *Nicotine Tob Res*. 2019; 21(3): 370-376.

Martinez-Pena Y Valenzuela I, Akaaboune M. The Metabolic Stability of the Nicotinic Acetylcholine Receptor at the Neuromuscular Junction. *Cells*. 2021; 10(2): 358.

Matta JA, Gu S, Davini WB, Bredt DS. Nicotinic acetylcholine receptor redux: Discovery of accessories opens therapeutic vistas. *Science*. 2021; 373(6556): eabg6539.

Rueda Ruzafa L, Cedillo JL, Hone AJ. Nicotinic Acetylcholine Receptor Involvement in Inflammatory Bowel Disease and Interactions with Gut Microbiota. *Int J Environ Res Public Health*. 2021; 18(3): 1189.

# Chapter 26

# Thyroglobulin (TG)

**Klaudia Dynarowicz**
**Dorota Bartusik-Aebisher***
**and David Aebisher**
Medical College of the University of Rzeszów, Poland

## Abstract

Thyroglobulin (TG) is a protein produced by the thyroid gland, more specifically by thyroid follicular cells. Its production is stimulated by intra-thyroid deficiency or excess of iodine and by the presence of immunoglobulins that stimulate the functioning of the thyroid gland. The serum TG level post-surgery reflects the amount of residual thyroid mass. Serum thyroglobulin measurement is essential in the diagnosis and follow-up of several thyroid disorders. An increasing TG concentration, when on a suppressive dose of thyroxine, indicates the recurrence of tumor or metastatic spread. Thyroglobulin assays are now in widespread use as a tumor marker for monitoring patients with differentiated thyroid carcinoma.

**Keywords:** thyroglobulin (TG), thyroid cancer, hypothyroidism, iodotyrosine, glycosylation, thyroid carcinoma

Thyroglobulin is a protein produced by the thyroid gland, more specifically by thyroid follicular cells. It is a glycoprotein that is synthesized by thyrocytes. Its production is stimulated by intra-thyroid deficiency or excess of iodine and

* Corresponding Author's Email: dbartusikaebisher@ur.edu.pl.

In: The Medical Biology Guide to Proteins
Editor: David Aebisher
ISBN: 979-8-88697-910-7

by the presence of immunoglobulins that stimulate the functioning of the thyroid gland. In a healthy patient, the value is between 3 and 40 ng / ml (Prpić, et al., 2018). The normal TG protein product is a homodimer containing 10% carbohydrates, with a molecular weight of 660,000. Like other secretory proteins, thyroglobulin is synthesized on ribosomes, glycosylated in the cisternae of the endoplasmic reticulum, translocated to the Golgi apparatus, and packaged in secretory vesicles, which discharge it from the apical surface into the lumen (Litonjua, et al., 1961).

Thyroglobulin can be produced by both good and normal thyroid cells, as well as cancerous cells. Therefore, this protein is also called a tumor marker. Elevated levels may indicate, for example, thyroid cancer or other diseases and conditions. Large deviations from the norm may indicate metastases or recurrences of the disease. It is currently known that over a quarter of patients diagnosed with thyroid cancer have an antibody against thyroglobulin. Occasionally, high levels of protein in the blood occur after partial thyroidectomy. Radioactive iodine is usually used when residual thyroid tissue needs to be removed. This is often when it reduces the risk of relapse or when more reliable data on the actual concentration of protein in the blood are obtained (Indrasena, 2017).

Thyroglobulin synthesizes peripheral thyroid hormones $T_3$ and $T_4$, which contain tyrosine residues iodinated with monoiodotyrosine and diiodotyrosine tyrosine oxidase, which are components of $T_3$ and $T_4$. Thyroglobulin as a neoplastic marker is a reliable marker in patients with differentiated thyroid cancer. In order to determine the level of thyroglobulin in the blood, a blood count is performed with the determination of this protein. TG can be secreted by differentiated cancer cells of the thyroid. This includes nearly two thirds of differentiated thyroid cancers (follicular neoplasms, all papillary carcinomas, Hurthle cell tumors of the thyroid), up to 50% of poorly differentiated and anaplastic carcinomas of thyroid and some medullary carcinomas. Structural changes in the Tg molecule or its precursors, inability to couple iodotyrosines, defective glycosylation, or abnormal transport through the membrane system of the cell could impair or substantially alter the synthesis of $T_4$ and $T_3$ and result in congenital goiter and various degrees of thyroid hypofunction (Coscia, et al., 2020). The half-life of thyroglobulin in the blood is approximately 65 hours. This concentration is proportional to the volume of thyroid tissue in the body. Due to the fact that the size of the normal thyroid gland ranges from 20-25 g, the reference standard of thyroglobulin concentration in the blood is from 20 to 25 ng / ml (Di Jeso, Arvan, 2016). The amount of iodine consumed by the body also affects the concentration of thyroglobilin. The concentration level also depends on gender. The reference norm in men is lower compared to women. The level of thyroglobulin is also influenced by antibodies and addictions, such as smoking. There are many causes of high thyroglobulin levels. Table 1 shows the most common causes of high thyroglobulin levels.

**Table 1.** The most common causes of high thyroglobulin levels

| Causes |
|---|
| Thyroid cancer |
| Poorly differentiated thyroid cancer |
| Benign thyroid nodules |
| Hypothyroidism |
| Graves disease |
| Iodine deficiency or excess |
| Cirrhosis |
| Anxieties |

The serum TG level post-surgery reflects the amount of residual thyroid mass. In the absence of metastatic disease, this reflects the size of the thyroid remnant left behind during surgery. Serum thyroglobulin measurement is essential in the diagnosis and follow-up of several thyroid disorders. This is especially the case in the management of differentiated thyroid cancer. Given the cellular specificity of thyreoglobulin, its detection in biopsy specimens provides proof of the thyroid origin of the tissue. In addition, measurements of serum thyreogobulin provide important information about the presence or absence of residual, recurrent, or metastatic disease in patients with differentiated thyroid cancer. Thyroglobulin assays are now in widespread use as a tumor marker for monitoring patients with differentiated thyroid carcinoma. In patients with papillary or follicular carcinoma and following total thyroidectomy and radioiodine ablation to remove the tumor, previously increased TG concentrations will reduce to very low or undetectable levels. An increasing TG concentration, when on a suppressive dose of thyroxine, indicates the recurrence of tumor or metastatic spread.

## References

Coscia F, Taler-Verčič A, Chang V T, Sinn L, O'Reilly F J, Izoré T, Renko M, Berger I, Rappsilber J, Turk D, Löwe J. The structure of human thyroglobulin. *Nature.* 2020 Feb;578(7796):627-630.

Di Jeso, B., Arvan, P. Thyroglobulin from molecular and cellular biology to clinical endocrinology. *Endocr. Rev.* 2016, 37, 2-36, doi:10.1210/er.2015-1090.

Indrasena, B.S. Use of thyroglobulin as a tumour marker. *World J. Biol. Chem.* 2017, 8, 81-85, doi:10.4331/wjbc.v8.i1.81.

Litonjua AD. Thyroglobulin. *Nature.* 1961 Jul 22;191:356-8.

Prpić, M., Franceschi, M., Romić, M., Jukić, T., Kusić, Z. Thyroglobulin as a tumor marker in differentiated thyroid cancer- clinical considerations. *Acta Clin. Croat.* 2018, 57, 518-527, doi:10.20471/acc.2018.57.03.16.

# Chapter 27

# Threonine

**Dominika Leś**
**Dorota Bartusik-Aebisher***
**and David Aebisher**
Medical College of the University of Rzeszów, Poland

## Abstract

Threonine is an organic chemical compound, an electrically neutral amino acid. Its full name is α-Amino-β-Hydroxybutyric Acid, and the chemical formula is C4H9NO3. It belongs to the exogenous amino acids, which means that the body is not able to produce it on its own, and at the same time it is essential and must be supplied with food.

**Keywords:** threonine, amino acids, protein

Threonine is an optically active amino acid (Read, et al., 2001). Threonine has many properties and has a huge impact on the proper functioning of the nervous system and the entire body. It is a building block of antibodies that are responsible for the immunity of the human body. In addition, it supports white blood cells in the fight against cancer cells and infections. Thanks to threonine, the body effectively resists the attacks of any microorganisms that could have a negative effect on its health. Threonine is also essential for the proper functioning of the nervous system. It has a positive effect on concentration, improves memory processes, enables more efficient

* Corresponding Author's Email: dbartusikaebisher@ur.edu.pl.

In: The Medical Biology Guide to Proteins
Editor: David Aebisher
ISBN: 979-8-88697-910-7

transmission of nerve information between neurons, reduces fatigue, and increases the resistance of neutrons to oxidative stress. The transmission of nervous information between neurons is responsible for reducing the effects of mental fatigue and normalizes the work of the central nervous system.

This exogenous amino acid is also involved in the production of collagen and elastin, elements important for the proper functioning of the skin, also ensures its proper hydration, protects it from draining too much water. Threonine is in the skin, and more precisely in the epidermis, the correct level delays the aging process, also makes the epidermis regenerate faster and is resistant to various damages. This is why this compound is often found in lotions and creams that are designed for dry and dehydrated skin. Such cosmetics ensure skin smoothness and elasticity and improve its level of hydration (Marrubini, et al., 2008). Threonine is also often added to shampoos and hair conditioners. Thanks to the regular use of cosmetics with the content of this amino acid, hair breakage is reduced, and static electricity is prevented.

Threonine is also involved in the synthesis of mucins in the gastrointestinal tract. Mucins are components of saliva and bile necessary for the digestion of food. Threonine is mainly absorbed in the upper part of the small intestine, and this is where it has a very strong protective effect. This amino acid is necessary for the formation of the mucus layer that covers the digestive tract, which is a natural barrier to digestive enzymes. In addition, threonine supports the proper metabolism of fats in the liver, thereby regulating the function of the liver (Moundras, et al., 1992). It can strengthen the enamel, regulate the work of the thyroid gland. Threonine also has a positive effect on the digestibility of nutrients, which is why it is often used in protein supplements aimed at athletes (Bartolomé, et al., 2022).

As mentioned earlier, threonine does not occur naturally in the human body, it is supplied with food. It is found in meat products, dairy products (up to 5%), whole grain products (up to 4.7%) and grains and legumes (up to 4%). The following products are good sources of this amino acid: pork, beef, beans, pumpkin seeds, sunflower seeds, peanuts, soybeans, tuna, mackerel, trout, milk, butter, cottage cheese, eggs (Reddy et al., 2007).

Depending on age, gender and weight, as well as physical activity, the requirement for threonine will vary. However, the average recommended dose of this amino acid is about 0.5 grams in the case of adults and as much as 3 grams in the case of children, whose protein requirements for building muscles are much higher.

Failure to meet the body's requirements for threonine can have a very bad effect on its functioning. The typical symptoms of deficiency of this amino acid include inhibition of protein metabolism, excessive emotional agitation, fatty liver, problems with the absorption of nutrients from food. The lack of threonine is synonymous with the lack of the rest of the minerals and vitamins, thanks to which the body can function properly. People suffering from depression also have a reduced level of threonine in the body. However, the excess of this component in the diet does not adversely affect the human body, it is not harmful to it.

Threonine is necessary for the proper functioning of the digestive tract and the entire body. It has a wide range of applications and numerous health-promoting properties. There is no doubt that threonine is one of the most important proteins in the human body, it is one of nearly 20 protein amino acids that should be supplied with food or additionally supplemented, because the body is unable to synthesize it on its own. The properly balanced diet allows one to provide the right amount of this amino acid.

## References

Bartolomé, M., Contento, A., M., Villaseñor, M., J., Ríos, Á. Innovative and versatile nanoplasmonic approach for the full sensing of proteinogenic aminacids in nutritional supplements. *Talanta.* 2022, 237.

Marrubini, G., Caccialanza, G., Massolini, G. Determination of glycine and threonine in topical dermatological preparations. *Journal of Pharmaceutical and Biomedical Analysis.* 2008, 47, 4-5, 716-722.

Moundras, C., Bercovici, D., Rémésy, C., Demigné, C. Influence of glucogenic amino acids on the hepatic metabolism of threonine. *Biochimica et Piophysica Acta (BBA) – General Subjects.* 1992, 1115, 3, 212-219.

Read, J., Brenner, S. Threonine. *Encyclopedia of Genetics.* 2001, 1965.

Reddy, M., M., Rudrabhatla, P., Rajasekharan, R. Importance of threonine residues in the regulation of peanut serice/threonine/tyrosine protein kinase activity. *Plant Science.* 2007, 172, 5, 1054-1059.

# Chapter 28

# Lysine

**Dominika Leś**
**Dorota Bartusik-Aebisher***
**and David Aebisher**
Medical College of the University of Rzeszów, Poland

## Abstract

Lysine (l-lysine; abbreviated name Lys, single letter abbreviation K) is an organic chemical compound belonging to the group of protein amino acids. For humans, it is an exogenous amino acid, i.e., it is not synthesized in the body and should be supplied with food.

**Keywords:** lysine, polar amino acids, protein

Lysine is a polar amino acid; it contains an alkaline side chain that is positively charged at the pH of the cell. It is a component of proteins that bind negatively charged nucleic acid molecules, such as histones (Murgola, 2001). Unlike other amino acids, it is stored in the body. Lysine in some foods can be the limiting amino acid, i.e., the amino acid that is present in the protein of a given product in the smallest amount in relation to the standard. This leads to the inefficient use of the remaining amino acids to build body proteins. Lysine is a ketogenic amino acid, which means that its metabolism produces products for ketogenesis. It is part of most proteins that build the cells of the body and plays a significant role in maintaining the proper functions of body proteins.

* Corresponding Author's Email: dbartusikaebisher@ur.edu.pl.

In: The Medical Biology Guide to Proteins
Editor: David Aebisher
ISBN: 979-8-88697-910-7

In addition, it participates in the metabolism of fats, conditioning the production of L-carnitine. Together with this amino acid, it strengthens the cardiovascular system and also prevents related development with plaque atherosclerosis. An increased need for lysine occurs in the case of difficult-to-heal wounds, osteoporosis, herpes simplex virus infection, adherence to a strict slimming diet, vegetarian diet and malnutrition.

Lysine has a protective function against the cardiovascular system. It helpin maintaining the correct structure of the artery walls and has a positive effect on the absorption of calcium in the body. Its correct level should be ensured by, inter alia, women at risk of osteoporosis. An adequate supply of lysine enables the effective use of the remaining amino acids for the construction of body proteins and the maintenance of their proper functions. It can affect the faster regeneration of muscles and the healing of wounds. Lysine and arginine use the same transport system in the body. Lysine just inhibits the production of arginine, an amino acid responsible for spreading the HSV herpes virus throughout the body (it is used as an auxiliary in case of infection with it). Research shows that the herpes simplex virus has a high demand for arginine. In the case of virus infection, lysine is used as an auxiliary. Its high level contributes to the reduction of arginine levels, and thus reduces the development of the herpes simplex virus (Strauss, et al., 2011).

Lysine is a building block of proteins in tendons, muscles and bones (Gorski, et al., 2021). It accelerates the regeneration of damaged tissues and the growth of muscle mass; it is extremely useful in physically active people and in the period of growth.

Lysine, participating in protein synthesis, is also involved in the production of enzymes, hormones and antibodies. It also contributes to the production of collagen, which is responsible for e.g., for the elasticity of the skin. As a result, the skin looks young, remains in good condition, and improves its elasticity. It also increases the absorption of calcium in the small intestine and is recommended for people exposed to its deficiency. It strengthens the immune system, increases immunity and alleviates the symptoms of colds and flu. Lysine can be used together with vitamin C, together they stimulate the body to produce antibodies, prevent infection and can also shorten its duration.

This amino acid has a very positive effect on mental concentration and adds energy. Providing the body with lysine can be helpful in fighting stress and anxiety. Lysine, in combination with other antioxidants, prevents the formation and the spread of cancer cells (Hao, et al., 2020). Lysine, as an essential chemical for the body, must be provided in the daily diet. It is found both in food products of both plant and animal origin. The products that contain it include: parsley, green peas, red beans, buckwheat, cocoa, whole grain bread, soy products, almonds, sesame seeds, chocolate, beef, chicken, turkey, fish, milk, eggs, gelatin, cheese, cottage cheese. According to data provided by health organizations, the need of an adult's body for lysine should be 30-35 mg for each kilogram of body weight, while in the case of children aged 13 to 18 it is 12 mg for each kilogram, and under the age of 13 it should be 10 mg for every kilogram. These values apply to healthy people in good condition. The demand may increase if the body is convalescing, is on a diet or is very active physically and mentally. The most common symptoms of lysine deficiency include, among others. fatigue, irritability, problems with concentration, decline in form, poor hair condition and hair loss, anemia, lack of appetite, herpes, dizziness, slow growth, muscle atrophy.

## References

Gorski, J. P., Franz, N. T., Pernoud, D., Keightley, A., Eyre, D. R., Oxford, J. T. A repeated triple lysine motif anchors complexes containing bone sialoprotein and the type XI collagen A1 chain involved in bone mineralization. *Journal of Biological Chemistry*. 2021, 296.

Hao, B., Sun, M., Zhang, M., Zhao, X., Zhao, L., Li, B., Zhai, L., Liu, P., Xu, J.-Y., Tan., M. Global chracterization of proteome and lysine methylome features in EZH2 wild-type and mutant lymphoma cel lines. *Journal of Proteomics*. 2020, 213.

Murgola, E. J. Lysine. *Brenner's Encyclopedia of Genetics* (Seocond Edition). 2001, 289.

Strauss, K. A., Brumbaugh, J., Duffy, A., Wardley, B., Robinson, D., Hendrickson, C., Tortorelli, S., Moser, A.B., Puffenberger, E. G., Rider, N. L., Morton, D. H. Safety, efficacy and physiological actions of a lysine-free, arginine-rich formula to treat glutaryl-CoA dehydrogenase deficiency: Focus on cerebral amino acid influx. *Molecular Genetics and Metabolism*. 2011, 104, 1-2, 93-106.

# Index

## F



## G



## H



## I



## L



## M



## N

## O



## P



## R



## S



## T

## V



## X

# Editor's Contact Information

*David Aebisher*
University of Rzeszow
Faculty of Medicine
Building G4
Al. Rejtana 16c, 35-959
Rzeszów, Poland
Email: daebisher@ur.edu.pl.